MARVELS OF MODERN SCIENCE

PAUL SEVERING

Edited by THEODORE WATERS

Marvels of Modern Science

Paul Severing

© 1st World Library – Literary Society, 2005
PO Box 2211
Fairfield, IA 52556
www.1stworldlibrary.org
First Edition

LCCN: 2004195658

Softcover ISBN: 1-4218-0479-4
Hardcover ISBN: 1-4218-0379-8
eBook ISBN: 1-4218-0579-0

Purchase *"Marvels of Modern Science"*
as a traditional bound book at:
www.1stWorldLibrary.org/purchase.asp?ISBN=1-4218-0479-4

1st World Library Literary Society is a nonprofit organization dedicated to promoting literacy by:

- Creating a free internet library accessible from any computer worldwide.
- Hosting writing competitions and offering book publishing scholarships.

Readers interested in supporting literacy through sponsorship, donations or membership please contact:
literacy@1stworldlibrary.org
Check us out at: www.1stworldlibrary.ORG
and start downloading free ebooks today.

Marvels of Modern Science
contributed by Tim, Ed & Rodney
in support of
1st World Library Literary Society

CONTENTS

INTRODUCTION 7

CHAPTER I 9
FLYING MACHINES

CHAPTER II 25
WIRELESS TELEGRAPHY

CHAPTER III 40
RADIUM

CHAPTER IV 49
MOVING PICTURES

CHAPTER V 59
SKY-SCRAPERS AND HOW THEY ARE BUILT

CHAPTER VI 69
OCEAN PALACES

CHAPTER VII 83
WONDERFUL CREATIONS IN PLANT LIFE

CHAPTER VIII 93
LATEST DISCOVERIES IN ARCHAEOLOGY

CHAPTER IX 105
GREAT TUNNELS OF THE WORLD

CHAPTER X 117
ELECTRICITY IN THE HOUSEHOLD

CHAPTER XI 125
HARNESSING THE WATER-FALL

CHAPTER XII 133
WONDERFUL WAR SHIPS

CHAPTER XIII 141
A TALK ON BIG GUNS

CHAPTER XIV 150
MYSTERY OF THE STARS

CHAPTER XV 161
CAN WE COMMUNICATE WITH OTHER WORLDS?

INTRODUCTION

The purpose of this little book is to give a general idea of a few of the great achievements of our time. Within such a limited space it was impossible to even mention thousands more of the great inventions and triumphs which mark the rushing progress of the world in the present century; therefore, only those subjects have been treated which appeal with more than passing interest to all. For instance, the flying machine is engaging the attention of the old, the young and the middle-aged, and soon the whole world will be on the wing. Radium, "the revealer," is opening the door to possibilities almost beyond human conception. Wireless Telegraphy is crossing thousands of miles of space with invisible feet and making the nations of the earth as one. 'Tis the same with the other subjects, - one and all are of vital, human interest, and are extremely attractive on account of their importance in the civilization of today. Mighty, sublime, wonderful, as have been the achievements of past science, as yet we are but on the verge of the continents of discovery. Where is the wizard who can tell what lies in the womb of time? Just as our conceptions of many things have been revolutionized in the past, those which we hold to-day of the cosmic processes may have to be remodeled in the future. The men of fifty years hence may laugh at the circumscribed knowledge of the present and shake their wise heads in contemplation of what they will term our crudities, and which we now call *progress*. Science is ever on the march and what is new to-day will be old to-morrow. We cannot go back, we must go forward, and although we can never reach finality in aught, we can improve on the *past* to enrich the *future*. If this volume creates an interest and arouses an enthusiasm in the ordinary men and women into whose hands it may come, and stimulates them to a study of the great events making for the enlightenment, progress and elevation of the race, it shall have fulfilled its mission and serve the purpose for which it was written.

CHAPTER I

FLYING MACHINES

Early Attempts at Flight - The Dirigible - Professor Langley's Experiment - The Wright Brothers - Count Zeppelin - Recent Aeroplane Records.

It is hard to determine when men first essayed the attempt to fly. In myth, legend and tradition we find allusions to aerial flight and from the very dawn of authentic history, philosophers, poets, and writers have made allusion to the subject, showing that the idea must have early taken root in the restless human heart. Aeschylus exclaims:

> "Oh, might I sit, sublime in air
> Where watery clouds the freezing snows prepare!"

Ariosto in his "Orlando Furioso" makes an English knight, whom he names Astolpho, fly to the banks of the Nile; nowadays the authors are trying to make their heroes fly to the North Pole.

Some will have it that the ancient world had a civilization much higher than the modern and was more advanced in knowledge. It is claimed that steam engines and electricity were common in Egypt thousands of years ago and that literature, science, art, and architecture flourished as never since. Certain it is that the Pyramids were for a long time the most solid "Skyscrapers" in the world.

Perhaps, after all, our boasted progress is but a case of going back to first principles, of history, or rather tradition repeating itself. The flying machine may not be as new as we think it is. At any rate the conception of it is old enough.

In the thirteenth century Roger Bacon, often called the "Father of Philosophy," maintained that the air could be navigated. He suggested a hollow globe of copper to be filled with "ethereal air or liquid fire," but he never tried to put his suggestion into practice. Father Vasson, a missionary at Canton, in a letter dated September 5, 1694, mentions a balloon that ascended on the occasion of the coronation of the Empress Fo-Kien in 1306, but he does not state where he got the information.

The balloon is the earliest form of air machine of which we have record. In 1767 a Dr. Black of Edinburgh suggested that a thin bladder could be made to ascend if filled with inflammable air, the name then given to hydrogen gas.

In 1782 Cavallo succeeded in sending up a soap bubble filled with such gas.

It was in the same year that the Montgolfier brothers of Annonay, near Lyons in France, conceived the idea of using hot air for lifting things into the air. They got this idea from watching the smoke curling up the chimney from the heat of the fire beneath.

In 1783 they constructed the first successful balloon of which we have any description. It was in the form of a round ball, 110 feet in circumference and, with the frame weighed 300 pounds. It was filled with 22,000 cubic feet of vapor. It rose to a height of 6,000 feet and proceeded almost 7,000 feet, when it gently descended. France went wild over the exhibition.

The first to risk their lives in the air were M. Pilatre de Rozier and the Marquis de Arlandes, who ascended over Paris in a hot-air balloon in November, 1783. They rose five hundred feet and traveled a distance of five miles in twenty-five minutes.

In the following December Messrs. Charles and Robert, also Frenchmen, ascended ten thousand feet and traveled twenty-seven miles in two hours.

The first balloon ascension in Great Britain was made by an experimenter named Tytler in 1784. A few months later Lunardi sailed over London.

In 1836 three Englishmen, Green, Mason and Holland, went from London to Germany, five hundred miles, in eighteen hours.

The greatest balloon exhibition up to then, indeed the greatest ever, as it has never been surpassed, was given by Glaisher and Coxwell, two Englishmen, near Wolverhampton, on September 5, 1862. They ascended to such an elevation that both lost the power of their limbs, and had not Coxwell opened the descending valve with his teeth, they would have ascended higher and probably lost their lives in the rarefied atmosphere, for there was no compressed oxygen then as now to inhale into their lungs. The last reckoning of which they were capable before Glaisher lost consciousness showed an elevation of twenty-nine thousand feet, but it is supposed that they ascended eight thousand feet higher before Coxwell was able to open the descending valve. In 1901 in the city of Berlin two Germans rose to a height of thirty-five thousand feet, but the two Englishmen of almost fifty years ago are still given credit for the highest ascent.

The largest balloon ever sent aloft was the "Giant" of M. Nadar, a Frenchman, which had a capacity of 215,000 cubic feet and required for a covering 22,000 yards of silk. It ascended from the Champ de Mars, Paris, in 1853, with fifteen passengers, all of whom came back safely.

The longest flight made in a balloon was that by Count de La Vaulx, 1193 miles in 1905.

A mammoth balloon was built in London by A. E. Gaudron. In 1908 with three other aeronauts Gaudron crossed from the Crystal Palace to the Belgian Coast at Ostend and then drifted over Northern Germany and was finally driven down by a snow storm at Mateki Derevni in Russia, having traveled 1,117 miles in 31-1/2 hours. The first attempt at constructing a dirigible balloon or airship was made by M. Giffard, a Frenchman, in 1852. The bag was spindle-shaped and 144 feet from point to point. Though it could be steered without drifting the motor was too weak to propel it. Giffard had many imitations in the spindle-shaped envelope construction, but it was a long time before any good results were obtained.

It was not until 1884 that M. Gaston Tissandier constructed a dirigible in any way worthy of the name.

It was operated by a motor driven by a bichromate of soda battery. The motor weighed 121 lbs. The cells held liquid enough to work for 2-1/2 hours, generating 1-1/3 horse power. The screw had two arms and was over nine feet in circumference. Tissandier made some successful flights.

The first dirigible balloon to return whence it started was that known as *La France*. This airship was also constructed in 1884. The designer was Commander Renard of the French Marine Corps assisted by Captain Krebs of the same service. The length of the envelope was 179 feet, its diameter 27-1/2 feet. The screw was in front instead of behind as in all others previously constructed. The motor which weighed 220-1/2 lbs. was driven by electricity and developed 8-1/2 horse power. The propeller was 24 feet in diameter and only made 46 revolutions to the minute. This was the first time electricity was used as a motor force, and mighty possibilities were conceived.

In 1901 a young Brazilian, Santos-Dumont, made a spectacular flight. M. Deutch, a Parisian millionaire, offered a prize of $20,000 for the first dirigible that would fly from the Parc d'Aerostat, encircle the Eiffel Tower and return to the starting point within thirty minutes, the distance of such flight being about nine miles. Dumont won the prize though he was some forty seconds over time. The length of his dirigible on this occasion was 108 feet, the diameter 19-1/2 feet. It had a 4-cylinder petroleum motor weighing 216 lbs., which generated 20 horse power. The screw was 13 feet in diameter and made three hundred revolutions to the minute.

From this time onward great progress was made in the

constructing of airships. Government officials and many others turned their attention to the work. Factories were put in operation in several countries of Europe and by the year 1905 the dirigible had been fairly well established. Zeppelin, Parseval, Lebaudy, Baidwin and Gross were crowding one another for honors. All had given good results, Zeppelin especially had performed some remarkable feats with his machines.

In the construction of the dirigible balloon great care must be taken to build a strong, as well as light framework and to suspend the car from it so that the weight will be equally distributed, and above all, so to contrive the gas contained that under no circumstances can it become tilted. There is great danger in the event of tilting that some of the stays suspending the car may snap and the construction fall to pieces in the air.

In deciding upon the shape of a dirigible balloon the chief consideration is to secure an end surface which presents the least possible resistance to the air and also to secure stability and equilibrium. Of course the motor, fuel and propellers are other considerations of vital importance.

The first experimenter on the size of wing surface necessary to sustain a man in the air, calculated from the proportion of weight and wing surface in birds, was Karl Meerwein of Baden. He calculated that a man weighing 200 lbs. would require 128 square feet. In 1781 he made a spindle-shaped apparatus presenting such a surface to the resistance of the air. It was collapsible on the middle and here the operator was fastened and lay horizontally with his face towards the earth working the collapsible wings by means of a transverse rod. It was not a success.

During the first half of the 19th Century there were many experiments with wing surfaces, none of which gave any promise. In fact it was not until 1865 that any advance was made, when Francis Wenham showed that the lifting power of a plane of great superficial area could be obtained by dividing the large plane into several parts arranged on tiers. This may be regarded as the germ of the modern aeroplane, the first glimmer of hope to filter through the darkness of experimentation until then. When Wenham's apparatus went against a strong wind it was only lifted up and thrown back. However, the idea gave thought to many others years afterwards.

In 1885 the brothers Lilienthal in Germany discovered the possibility of driving curved aeroplanes against the wind. Otto Lilienthal held that it was necessary to begin with "sailing" flight and first of all that the art of balancing in the air must be learned by practical experiments. He made several flights of the kind now known as *gliding*. From a height of 100 feet he glided a distance of 700 feet and found he could deflect his flight from left to right by moving his legs which were hanging freely from the seat. He attached a light motor weighing only 96 lbs. and generating 2-1/2 horse power. To sustain the weight he had to increase the size of his planes.

Unfortunately this pioneer in modern aviation was killed in an experiment, but he left much data behind which has helped others. His was the first actual flyer which demonstrated the elementary laws governing real flight and blazed the way for the successful experiments of the present time. His example made the gliding machine a continuous performance until real practical aerial flight was achieved.

As far back as 1894 Maxim built a giant aeroplane but it was too cumbersome to be operated.

In America the wonderful work of Professor Langley of the Smithsonian Institution with his aerodromes attracted worldwide attention. Langley was the great originator of the science of aerodynamics on this side of the water. Langley studied from artificial birds which he had constructed and kept almost constantly before him.

To Langley, Chanute, Herring and Manly, America owes much in the way of aeronautics before the Wrights entered the field. The Wrights have given the greatest impetus to modern aviation. They entered the field in 1900 and immediately achieved greater results than any of their predecessors. They followed the idea of Lilienthal to a certain extent. They made gliders in which the aviator had a horizontal position and they used twice as great a lifting surface as that hitherto employed. The flights of their first motor machine was made December 17, 1903, at Kitty Hawk, N.C. In 1904 with a new machine they resumed experiments at their home near Dayton, O. In September of that year they succeeded in changing the course from one dead against the wind to a curved path where cross currents must be encountered, and made many circular flights. During 1906 they rested for a while from practical flight, perfecting plans for the future. In the beginning of September, 1908, Orville Wright made an aeroplane flight of one hour, and a few days later stayed up one hour and fourteen minutes. Wilbur Wright went to France and began a series of remarkable flights taking up passengers. On December 31, of that year, he startled the world by making the record flight of two hours and nineteen minutes.

It was on Sept. 13, 1906, that Santos-Dumont made the first officially recorded European aeroplane flight, leaving the ground for a distance of 12 yards. On November 12, of same year, he remained in the air for 21 seconds and traveled a distance of 230 yards. These feats caused a great sensation at the time.

While the Wrights were achieving fame for America, Henri Farman was busy in England. On October 26, 1907, he flew 820 yards in 52-1/2 seconds. On July 6, 1908, he remained in the air for 20-1/2 minutes. On October 31, same year, in France, he flew from Chalons to Rheims, a distance of sixteen miles, in twenty minutes.

The year 1909 witnessed mighty strides in the field of aviation. Thousands of flights were made, many of which exceeded the most sanguine anticipations. On July 13, Bleriot flew from Etampes to Chevilly, 26 miles, in 44 minutes and 30 seconds, and on July 25 he made the first flight across the British Channel, 32 miles, in 37 minutes. Orville Wright made several sensational flights in his biplane around Berlin, while his brother Wilbur delighted New Yorkers by circling the Statue of Liberty and flying up the Hudson from Governor's Island to Grant's Tomb and return, a distance of 21 miles, in 33 minutes and 33 seconds during the Hudson-Fulton Celebration. On November 20 Louis Paulhan, in a biplane, flew from Mourmelon to Chalons, France, and return, 37 miles in 55 minutes, rising to a height of 1000 feet.

The dirigible airship was also much in evidence during 1909, Zeppelin, especially, performing some remarkable feats. The Zeppelin V., subsequently re-numbered No. 1, of the rigid type, 446 feet long, diameter 42-1/2 feet and capacity 536,000 cubic feet, on March 29, rose to a height of 3,280, and on April 1, started with a crew of nine passengers from Frederickshafen to Munich. In a 35 mile gale it was carried beyond Munich, but Zeppelin succeeded in coming to anchor. Other Zeppelin balloons made remarkable voyages during the year. But the latest achievements (1910) of the old German aeronaut have put all previous records into the shade and electrified the whole world. His new passenger airship, the *Deutschland*, on June 22, made a 300 mile trip from Frederickshafen to Dusseldorf in 9 hours, carrying 20 passengers. This was at the rate of 33.33 miles per hour. During one hour of the journey a speed of 43-1/2 miles was averaged. The passengers were carried in a mahogany finished cabin and had all the comforts of a Pullman car, but most significant fact of all, the trip was made on schedule and with all regularity of an express train.

Two days later Zeppelin eclipsed his own record air voyage when his vessel carried 32 passengers, ten of whom were women, in a 100 mile trip from Dusseldorf to Essen, Dortmund and Bochum and back. At one time on this occasion while traveling with the wind the airship made a speed of 56-1/2 miles. It passed through a heavy shower and forced its way against a strong headwind without difficulty. The passengers were all delighted with the new mode of travel, which was very comfortable. This last dirigible masterpiece of Zeppelin may be styled the leviathan of the air. It is 485 feet long with a total lifting power of 44,000 lbs. It has three motors which total 330 horse power and it drives at an average speed of about 33 miles an hour. A regular passenger service has been established and tickets are selling at $50.

The present year can also boast some great aeroplane records, notably by Curtiss and Hamilton in America and Farman and Paulhan in Europe. Curtiss flew from Albany to New York, a distance of 137 miles, at an average speed of 55 miles an hour and Hamilton flew from New York to Philadelphia and return. The first night flight of a dirigible over New York City was made by Charles Goodale on July 19. He flew from Palisades Park on the Hudson and return.

From a scientific toy the Flying Machine has been developed and perfected into a practical means of locomotion. It bids fair at no distant date to revolutionize the transit of the world. No other art has ever made such progress in its early stages and every day witnesses an improvement.

The air, though invisible to the eye, has mass and therefore offers resistance to all moving bodies. Therefore air-mass and air resistance are the first principles to be taken into consideration in the construction of an aeroplane. It must be built so that the air-mass will sustain it and the motor, and the motor must be of sufficient power to overcome the air resistance.

A ship ploughing through the waves presents the line of least resistance to the water and so is shaped somewhat like a fish, the natural denizen of that element. It is different with the aeroplane. In the intangible domain it essays to overcome, there must be a sufficient surface to compress a certain volume of air to sustain the weight of the machinery.

The surfaces in regard to size, shape, curvature, bracing and material, are all important. A great deal depends upon the curve of the surfaces. Two machines may have the same extent of surface and develop the same rate of speed, yet one may have a much greater lifting power than the other, provided it has a more efficient curve to its surface. Many people have a fallacious idea that the surfaces of an aeroplane are planes and this doubt less arises from the word itself. However, the last syllable in *aeroplane* has nothing whatever to do with a flat surface. It is derived from the Greek *planos*, wandering, therefore the entire word signifies an air wanderer.

The surfaces are really aero curves arched in the rear of the front edge, thus allowing the supporting surface of the aeroplane in passing forward with its backward side set at an angle to the direction of its motion, to act upon the air in such a way as to tend to compress it on the under side.

After the surfaces come the rudders in importance. It is of vital consequence that the machine be balanced by the operator. In the present method of balancing an aeroplane the idea in mind is to raise the lower side of the machine and make the higher side lower in order that it can be quickly righted when it tips to one side from a gust of wind, or when making angle at a sudden turn. To accomplish this, two methods can be employed. 1. Changing the form of the wing. 2. Using separate surfaces. One side can be made to lift more than the other by giving it a greater curve or extending the extremity.

In balancing by means of separate surfaces, which can be turned up or down on each side of the machine, the horizontal balancing rudders are so connected that they will work in an opposite direction - while one is turned to lift one side, the other will act to lower the other side so as to strike an even balance.

The motors and propellers next claim attention. It is the motor that makes aviation possible. It was owing in a very large measure to the introduction of the petrol motor that progress became rapid. Hitherto many had laid the blame of everything on the motor. They had said, - "give us a light and powerful engine and we will show you how to fly."

The first very light engine to be available was the *Antoinette*, built by Leon Levavasseur in France. It enabled Santos-Dumont to make his first public successful flights. Nearly all aeroplanes follow the same general principles of construction. Of course a good deal depends upon the form of aeroplane - whether a monoplane or a biplane. As these two forms are the chief ones, as yet, of heavier than-air machines, it would be well to understand them. The monoplane has single large surfaces like the wings of a bird, the biplane has two large surfaces braced together one over the other. At the present writing a triplane has been introduced into the domain of American aviation by an English aeronaut. Doubtless as the science progresses many other variations will appear in the field. Most machines, though fashioned on similar lines, possess universal features. For instance, the Wright biplane is characterized by warping wing tips and seams of heavy construction, while the surfaces of the Herring-Curtiss machine, are slight and it looks very light and buoyant as if well suited to its element. The Voisin biplane is fashioned after the manner of a box kite and therefore presents vertical surfaces to the air. Farman's machine has no vertical surfaces, but there are hinged wing tips to the outer rear-edges of its surfaces, for use in turning and balancing. He also has a combination of wheels and skids or runners for starting and landing.

The position to be occupied by the operator also influences the construction. Some sit on top of the machine, others underneath. In the *Antoinette*, Latham sits up in a sort of cockpit on the top. Bleriot sits far beneath his machine. In the latest construction of Santos-Dumont, the *Demoiselle*, the aviator sits on the top.

Aeroplanes have been constructed for the most part in Europe, especially in France. There may be said to be only one factory in America, that of Herring-Curtiss, at Hammondsport, N.Y., as the Wright place at Dayton is very small and only turns out motors and experimenting machines, and cannot be called a regular factory. The Wright machines are now manufactured by a French syndicate. It is said that the Wrights will have an American factory at work in a short time. The French-made aeroplanes have given good satisfaction. These machines cost from $4,000 to $5,000, and generally have three cylinder motors developing from 25 to 35 horse power.

The latest model of Bleriot known as No. 12 has beaten the time record of Glenn Curtiss' biplane with its 60 horse power motor. The Farman machine or the model in which he made the world's duration record in his three hour and sixteen minutes flight at Rheims, is one of the best as well as the cheapest of the French makes. Without the motor it cost but $1,200. It has a surface twenty-five meters square, is eight meters long and seven-and-a-half meters wide, weighs 140 kilos, and has a motor which develops from 25 to 50 horse power.

The Wright machines cost $6,000. They have four cylinder motors of 30 horse power, are 12-1/2 meters long, 9 meters wide and have a surface of 30 square meters. They weigh 400 kilos. In this country they cost $7,500 exclusive of the duty on foreign manufacture.

The impetus being given to aviation at the present time by the prizes offered is spurring the men-birds to their best efforts.

It is prophesied that the aeroplane will yet attain a speed of 300 miles an hour. The quickest travel yet attained by man has been at the rate of 127 miles an hour. That was accomplished by Marriott in a racing automobile at Ormond Beach in 1906, when he went one mile in 28 1-5 seconds. It is doubtful, however, were it possible to achieve a rate of 300 miles an hour, that any human being could resist the air pressure at such a velocity.

At any rate there can be no question as to the aeroplane attaining a much greater speed than at present. That it will be useful there can be little doubt. It is no longer a scientific toy in the hands of amateurs, but a practical machine which is bound to contribute much to the progress of the world. Of course, as a mode of transportation it is not in the same class with the dirigible, but it can be made to serve many other purposes. As an agent in time of war it would be more important than fort or warship.

The experiments of Curtiss, made a short time ago

over Lake Keuka at Hammondsport, N.Y., prove what a mighty factor would have to be reckoned with in the martial aeroplane. Curtiss without any practice at all hit a mimic battle ship fifteen times out of twenty-two shots. His experiment has convinced the military and naval authorities of this country that the aeroplane and the aerial torpedo constitute a new danger against which there is no existing protection. Aerial offensive and defensive strategy is now a problem which demands the attention of nations.

CHAPTER II

WIRELESS TELEGRAPHY

Primitive Signalling - Principles of Wireless Telegraphy - Ether Vibrations - Wireless Apparatus - The Marconi System.

At a very early stage in the world's history, man found it necessary to be able to communicate with places at a distance by means of signals. Fire was the first agent employed for the purpose. On hill-tops or other eminences, what were known as beacon fires were kindled and owing to their elevation these could be seen for a considerable distance throughout the surrounding country. These primitive signals could be passed on from one point to another, until a large region could be covered and many people brought into communication with one another. These fires expressed a language of their own, which the observers could readily interpret. For a long time they were the only method used for signalling. Indeed in many backward localities and in some of the outlying islands and among savage tribes the custom still prevails. The bushmen of Australia at night time build fires outside their huts or kraals to attract the attention of their followers.

Even in enlightened Ireland the kindling of beacon fires is still observed among the people of backward districts especially on May Eve and the festival of midsummer. On these occasions bonfires are lit on almost every hillside throughout that country. This custom has been handed down from the days of the Druids.

For a long time fires continued to be the mode of signalling, but as this way could only be used in the night, it was found necessary to adopt some method that would answer the purpose in daytime; hence signal towers were erected from which flags were waved and various devices displayed. Flags answered the purposes so very well that they came into general use. In course of time they were adopted by the army, navy and merchant marine and a regular code established, as at the present time.

The railroad introduced the semaphore as a signal, and field tactics the heliograph or reflecting mirror which, however, is only of service when there is a strong sunlight.

Then came the electric telegraph which not only revolutionized all forms of signalling but almost annihilated distance. Messages and all sorts of communications could be flashed over the wires in a few minutes and when a cable was laid under the ocean, continent could converse with continent as if they were next door neighbors.

The men who first enabled us to talk over a wire certainly deserve our gratitude, all succeeding generations are their debtors. To the man who enabled us to talk to long distances without a wire at all it would seem we owe a still greater debt. But who is this man around whose brow we should twine the laurel wreath, to the altar of whose genius we should carry frankincense and myrrh?

This is a question which does not admit of an answer, for to no one man alone do we owe wireless telegraphy, though Hertz was the first to discover the waves which make it possible. However, it is to the men whose indefatigable labors and genius made the electric telegraph a reality, that we also owe wireless telegraphy as we have it at present, for the latter may be considered in many respects the resultant of the former, though both are different in medium.

Radio or wireless telegraphy in principle is as old as mankind. Adam delivered the first wireless when on awakening in the Garden of Eden he discovered Eve and addressed her in the vernacular of Paradise in that famous sentence which translated in English reads both ways the same, - "Madam, I'm Adam." The oral words issuing from his lips created a sound wave which the medium of the air conveyed to the tympanum of the partner of his joys and the cause of his sorrows.

When one person speaks to another the speaker causes certain vibrations in the air and these so stimulate the hearing apparatus that a series of nerve impulses are conveyed to the sensorium where the meaning of these signals is unconsciously interpreted.

In wireless telegraphy the sender causes vibrations not in the air but in that all-pervading impalpable substance which fills all space and which we call the ether. These vibrations can reach out to a great distance and are capable of so affecting a receiving apparatus that signals are made, the movements of which can be interpreted into a distinct meaning and consequently into the messages of language.

Let us briefly consider the underlying principles at work. When we cast a stone into a pool of water we observe that it produces a series of ripples which grow fainter and fainter the farther they recede from the centre, the initial point of the disturbance, until they fade altogether in the surrounding expanse of water. The succession of these ripples is what is known as *wave* motion.

When the clapper strikes the lip of a bell it produces a sound and sends a tremor out upon the air. The vibrations thus made are air waves.

In the first of these cases the medium communicating the ripple or wavelet is the water. In the second case the medium which sustains the tremor and communicates the vibrations is the air.

Let us now take the case of a third medium, the substance of which puzzled the philosophers of ancient time and still continues to puzzle the scientists of the present. This is the ether, that attenuated fluid which fills all inter-stellar space and all space in masses and between molecules and atoms not otherwise occupied by gross matter. When a lamp is lit the light radiates from it in all directions in a wave motion. That which transmits the light, the medium, is ether. By this means energy is conveyed from the sun to the earth, and scientists have calculated the speed of the ether vibrations called light at 186,400 miles per second. Thus a beam of light can travel from the sun to the earth, a distance of between 92,000,000 and 95,000,000 miles (according to season), in a little over eight minutes.

The fire messages sent by the ancients from hill to hill were ether vibrations. The greater the fires, the greater were the vibrations and consequently they carried farther to the receiver, which was the eye. If a signal is to be sent a great distance by light the source of that light must be correspondingly powerful in order to disturb the ether sufficiently. The same principle holds good in wireless telegraphy. If we wish to communicate to a great distance the ether must be disturbed in proportion to the distance. The vibrations that produce light are not sufficient in intensity to affect the ether in such a way that signals can be carried to a distance. Other disturbances, however, can be made in the ether, stronger than those which create light. If we charge a wire with an electric current and place a magnetic needle near it we find it moves the needle from one position to another. This effect is called an electro-magnetic disturbance in the ether. Again when we charge an insulated body with electricity we find that it attracts any light substance indicating a material disturbance in the ether. This is described as an electro-static disturbance or effect and it is upon this that wireless telegraphy depends for its operations.

The late German physicist, Dr. Heinrich Hertz, Ph.D., was the first to detect electrical waves in the ether. He set up the waves in the ether by means of an electrical discharge from an induction coil. To do this he employed a very simple means. He procured a short length of wire with a brass knob at either end and bent around so as to form an almost complete circle leaving only a small air gap between the knobs. Each time there was a spark discharge from the induction coil, the experimenter found that a small electric spark also generated between the knobs of the wire loop, thus showing that electric waves were projected through the ether. This discovery suggested to scientists that such electric waves might be used as a means of transmitting signals to a distance through the medium of the ether without connecting wires.

When Hertz discovered that electric waves crossed space he unconsciously became the father of the modern system of radio-telegraphy, and though he did not live to put or see any practical results from his wonderful discovery, to him in a large measure should be accorded the honor of blazoning the way for many of the intellectual giants who came after him. Of course those who went before him, who discovered the principles of the electric telegraph made it possible for the Hertzian waves to be utilized in wireless.

It is easy to understand the wonders of wireless telegraphy when we consider that electric waves transverse space in exactly the same manner as light waves. When energy is transmitted with finite velocity we can think of its transference only in two ways: first by the actual transference of matter as when a stone is hurled from one place to another; second, by the propagation of energy from point to point through a medium which fills the space between two bodies. The body sending out energy disturbs the medium contiguous to it, which disturbance is communicated to adjacent parts of the medium and so the movement is propagated outward from the sending body through the medium until some other body is affected.

A vibrating body sets up vibrations in another body, as for instance, when one tuning fork responds to the vibrations of another when both have the same note or are in tune.

The transmission of messages by wireless telegraphy is effected in a similar way. The apparatus at the sending station sends out waves of a certain period through the ether and these waves are detected at the receiving station, by apparatus attuned to this wave length or period.

The term electric radiation was first employed by Hertz to designate waves emitted by a Leyden jar or oscillator system of an induction coil, but since that time these radiations have been known as Hertzian waves. These waves are the underlying principles in wireless telegraphy.

It was found that certain metal filings offered great resistance to the passage of an electric current through them but that this resistance was very materially reduced when electric waves fell upon the filings and remained so until the filings were shaken, thus giving time for the fact to be observed in an ordinary telegraphic instrument.

The tube of filings through which the electric current is made to pass in wireless telegraphy is called a coherer signifying that the filings cohere or cling together under the influence of the electric waves. Almost any metal will do for the filings but it is found that a combination of ninety per cent. nickel and ten per cent. silver answers the purpose best.

The tube of the coherer is generally of glass but any insulating substance will do; a wire enters at each end and is attached to little blocks of metal which are separated by a very small space. It is into this space the filings are loosely filled.

Another form of coherer consists of a glass tube with small carbon blocks or plugs attached to the ends of the wires and instead of the metal filings there is a globule of mercury between the plugs. When electric waves fall upon this coherer, the mercury coheres to the carbon blocks, and thus forms a bridge for the battery current.

Marconi and several others have from time to time invented many other kinds of detectors for the electrical waves. Nearly all have to serve the same purpose, viz., to close a local battery circuit when the electric waves fall upon the detector.

There are other inventions on which the action is the reverse. These are called anti-coherers. One of the best known of these is a tube arranged in a somewhat similar manner to the filings tube but with two small blocks of tin, between which is placed a paste made up of alcohol, tin filings and lead oxide. In its normal state the paste allows the battery current to get across from one block to another, but when electric waves touch it a chemical action is produced which immediately breaks down the bridge and stops the electric waves, the paste resumes its normal condition and allows the battery current to pass again. Therefore by this arrangement the signals are made by a sudden breaking and making of the battery circuit.

Then there is the magnetic detector. This is not so easy of explanation. When we take a piece of soft iron and continuously revolve it in front of a permanent magnet, the magnetic poles of the soft iron piece will keep changing their position at each half revolution. It requires a little time to effect this magnetic change which makes it appear as if a certain amount of resistance was being made against it. (If electric waves are allowed to fall upon the iron, resistance is completely eliminated, and the magnetic poles can change places instantly as it revolves.)

From this we see that if we have a quickly changing magnetic field it will induce or set up an electric current in a neighboring coil of wire. In this way we can detect the changes in the magnetic field, for we can place a telephone receiver in connection with the coil of wire.

In a modern wireless receiver of this kind it is found more convenient to replace the revolving iron piece by an endless band of soft iron wire. This band is kept passing in front of a permanent magnet, the magnetism of the wire tending to change as it passes from one pole to the other. This change takes place suddenly when the electric waves form the transmitting station, fall upon the receiving aerial conductor and are conducted round the moving wire, and as the band is passing through a coil of insulated wire attached to a telephone receiver, this sudden change in the magnetic field induces an electric current in the surrounding coil and the operator hears a sound in the telephone at his ear. The Morse code may thus be signalled from the distant transmitter.

There are various systems of wireless telegraphy for the most part called after the scientists who developed or perfected them. Probably the foremost as well as the best known is that which bears the name of Marconi. A popular fallacy makes Marconi the discoverer of the wireless method. Marconi was the first to put the system on a commercial footing or business basis and to lead the way for its coming to the front as a mighty factor in modern progress. Of course, also, the honor of several useful inventions and additions to wireless apparatus must be given him. He started experimenting as far back as 1895 when but a mere boy. In the beginning he employed the induction coil, Morse telegraph key, batteries, and vertical wire for the transmission of signals, and for their reception the usual filings coherer of nickel with a very small percentage of silver, a telegraph relay, batteries and a vertical wire. In the Marconi system of the present time there are many forms of coherers, also the magnetic detector and other variations of the original apparatus. Other systems more or less prominent are the Lodge-Muirhead of England, Braun-Siemens of Germany and those of DeForest and Fessenden of America. The electrolytic detector with the paste between the tin blocks belongs to the system of DeForest. Besides these the names of Popoff, Jackson, Armstrong, Orling, Lepel, and Poulsen stand high in the wireless world.

A serious drawback to the operations of wireless lies in the fact that the stations are liable to get mixed up and some one intercept the messages intended for another, but this is being overcome by the adoption of a special system of wave lengths for the different wireless stations and by the use of improved apparatus.

In the early days it was quite a common occurrence for the receivers of one system to reply to the transmitters of a rival system. There was an all-round mix-up and consequently the efficiency of wireless for practical purposes was for a good while looked upon with more or less suspicion. But as knowledge of wave motions developed and the laws of governing them were better understood, the receiver was "tuned" to respond to the transmitter, that is, the transmitter was made to set up a definite rate of vibrations in the ether and the receiver made to respond to this rate, just like two tuning forks sounding the same note.

In order to set up as energetic electric waves as possible many methods have been devised at the transmitting stations. In some methods a wire is attached to one of the two metal spheres between which the electric charge takes place and is carried up into the air for a great height, while to the second sphere another wire is connected and which leads into the earth. Another method is to support a regular network of wires from strong steel towers built to a height of two hundred feet or more.

Long distance transmission by wireless was only made possible by grounding one of the conductors in the transmitter. The Hertzian waves were provided without any earth connection and radiated into space in all directions, rapidly losing force like the disappearing ripples on a pond, whereas those set up by a grounded transmitter with the receiving instrument similarly connected to earth, keep within the immediate neighborhood of the earth.

For instance up to about two hundred miles a storage battery and induction coil are sufficient to produce the necessary ether disturbance, but when a greater distance is to be spanned an engine and a dynamo are necessary to supply energy for the electric waves.

In the most recent Marconi transmitter the current

produced is no longer in the form of intermittent sparks, but is a true alternating current, which in general continues uniformly as long as the key is pressed down.

There is no longer any question that wireless telegraphy is here to stay. It has passed the juvenile stage and is fast approaching a lusty adolescence which promises to be a source of great strength to the commerce of the world. Already it has accomplished much for its age. It has saved so many lives at sea that its installation is no longer regarded as a scientific luxury but a practical necessity on every passenger vessel. Practically every steamer in American waters is equipped with a wireless station. Even freight boats and tugs are up-to-date in this respect. Every ship in the American navy, including colliers and revenue cutters, carries wireless operators. So important indeed is it considered in the Navy department that a line of shore stations have been constructed from Maine on the Atlantic to Alaska on the Pacific.

In a remarkably short interval wireless has come to exercise an important function in the marine service. Through the shore stations of the commercial companies, press despatches, storm warnings, weather reports and other items of interest are regularly transmitted to ships at sea. Captains keep in touch with one another and with the home office; wrecks, derelicts and storms are reported. Every operator sends out regular reports daily, so that the home office can tell the exact position of the vessel. If she is too far from land on the one side to be reached by wireless she is near enough on the other to come within the sphere of its operations.

Weather has no effect on wireless, therefore the question of meteorology does not come into consideration. Fogs, rains, torrents, tempests, snowstorms, winds, thunder, lightning or any aerial disturbance whatsoever cannot militate against the sending or receiving of wireless messages as the ether permeates them all.

Submarine and land telegraphy used to look on wireless, the youngest sister, as the Cinderella of their name, but she has surpassed both and captured the honors of the family. It was in 1898 that Marconi made his first remarkable success in sending messages from England to France. The English station was at South Foreland and the French near Boulogne. The distance was thirty-two miles across the British channel. This telegraphic communication without wires was considered a wonderful feat at the time and excited much interest.

During the following year Marconi had so much improved his first apparatus that he was able to send out waves detected by receivers up to the one hundred mile limit.

In 1900 communication was established between the Isle of Wight and the Lizard in Cornwall, a distance of two hundred miles.

Up to this time the only appliances employed were induction coils giving a ten or twenty inch spark. Marconi and others perceived the necessity of employing greater force to penetrate the ether in order to generate stronger electrical waves. Oil and steam engines and other appliances were called into use to create high frequency currents and those necessitated the erection of large power stations. Several were erected at advantageous points and the wireless system was fairly established as a new agent of communication.

In December, 1901, at St. John's, Newfoundland, Marconi by means of kites and balloons set up a temporary aerial wire in the hope of being able to receive a signal from the English station in Cornwall. He had made an arrangement with Poldhu station that on a certain date and at a fixed hour they should attempt the signal. The letter S, which in the Morse code consists of three successive dots, was chosen. Marconi feverishly awaited results. True enough on the day and at the time agreed upon the three dots were clicked off, the first signal from Europe to the American continent. Marconi with much difficulty set up other aerial wires and indubitably established the fact that it was possible to send electric waves across the Atlantic. He found, however, that waves in order to traverse three thousand miles and retain sufficient energy on their arrival to affect a telephonic wave-detecting device must be generated by no inordinate power.

These experiments proved that if stations were erected of sufficient power transatlantic wireless could be successfully carried on. They gave an impetus to the erection of such stations.

On December 21, 1902, from a station at Glace Bay, Nova Scotia, Marconi sent the first message by wireless to England announcing success to his colleagues.

The following January from Wellsfleet, Cape Cod,

President Roosevelt sent a congratulatory message to King Edward. The electric waves conveying this message traveled 3,000 miles over the Atlantic following round an arc of forty-five degrees of the earth on a great circle, and were received telephonically, by the Marconi magnetic receiver at Poldhu.

Most ships are provided with syntonic receivers which are tuned to long distance transmitters, and are capable of receiving messages up to distances of 3,000 miles or more. Wireless communication between Europe and America is no longer a possibility but an accomplishment, though as yet the system has not been put on a general business basis. [Footnote: As we go to press a new record has been established in wireless transmission. Marconi, in the Argentine Republic, near Buenos Ayres, has received messages from the station at Clifden, County Galway, Ireland, a distance of 5,600 miles. The best previous record was made when the United States battleship *Tennessee* in 1909 picked up a message from San Francisco when 4,580 miles distant.]

CHAPTER III

RADIUM

Experiments of Becquerel - Work of the Curies - Discovery of Radium - Enormous Energy - Various Uses.

Early in 1896 just a few months after Roentgen had startled the scientific world by the announcement of the discovery of the X-rays, Professor Henri Becquerel of the Natural History Museum in Paris announced another discovery which, if not as mysterious, was more puzzling and still continues a puzzle to a great degree to the present time. Studying the action of the salts of a rare and very heavy mineral called uranium Becquerel observed that their substances give off an invisible radiation which, like the Roentgen rays, traverse metals and other bodies opaque to light, as well as glass and other transparent substances. Like most of the great discoveries it was the result of accident. Becquerel had no idea of such radiations, had never thought of their possibility.

In the early days of the Roentgen rays there were many facts which suggested that phosphorescence had something to do with the production of these rays It then occurred to several French physicists that X-rays might be produced if phosphorescent substances were exposed to sunlight. Becquerel began to experiment with a view to testing this supposition. He placed uranium on a photographic plate which had first been wrapped in black paper in order to screen it from the light. After this plate had remained in the bright sunlight for several hours it was removed from the paper covering and developed. A slight trace of photographic action was found at those parts of the plate directly beneath the uranium just as Becquerel had expected. From this it appeared evident that rays of some kind were being produced that were capable of passing through black paper. Since the X-rays were then the only ones known to possess the power to penetrate opaque substances it seemed as though the problem of producing X-rays by sunlight was solved. Then came the fortunate accident. After several plates had been prepared for exposure to sunlight a severe storm arose and the experiments had to be abandoned for the time being. At the end of several days work was again resumed, but the plates had been lying so long in the darkroom that they were deemed almost valueless and it was thought that there would not be much use in trying to use them. Becquerel was about to throw them away, but on second consideration thinking that some action might have possibly taken place in the dark, he resolved to try them. He developed them and the result was that he obtained better pictures than ever before. The exposure to sunlight which had been regarded as essential to the success of the former experiments had really nothing at all to do with the matter, the essential thing was the presence of uranium and the photographic effects were not due to X-rays but to the rays or emanations which Becquerel had thus discovered and which bear his name.

There were many tedious and difficult steps to take before even our present knowledge, incomplete as it is, could be reached. However, Becquerel's fortunate accident of the plate developing was the beginning of the long series of experiments which led to the discovery of radium which already has revolutionized some of the most fundamental conceptions of physics and chemistry.

It is remarkable that we owe the discovery of this wonderful element to a woman, Mme. Sklodowska Curie, the wife of a French professor and physicist. Mme. Curie began her work in 1897 with a systematic study of several minerals containing uranium and thorium and soon discovered the remarkable fact that there was some agent present more strongly radio-active than the metal uranium itself. She set herself the task of finding out this agent and in conjunction with her husband, Professor Pierre Curie, made many tests and experiments. Finally in the ore of pitchblende they found not only one but three substances highly radio-active. Pitchblende or uraninite is an intensely black mineral of a specific gravity of 9.5 and is found in commercial quantities in Bohemia, Cornwall in England and some other localities. It contains lead sulphide, lime silica, and other bodies.

To the radio-active substance which accompanied the bismuth extracted from pitchblende the Curies gave the name *Polonium*. To that which accompanied barium taken from the same ore they called *Radium* and to the substance which was found among the rare earths of the pitchblende Debierne gave the name *Actinium*.

None of these elements have been isolated, that is to say, separated in a pure state from the accompanying ore. Therefore, *pure radium* is a misnomer, though we often hear the term used. [Footnote: Since the above was written Madame Curie has announced to the Paris Academy of Sciences that she has succeeded in obtaining pure radium. In conjunction with Professor Debierne she treated a decegramme of bromide of radium by electrolytic process, getting an amalgam from which was extracted the metallic radium by distillation.] All that has been obtained is some one of its simpler salts or compounds and until recently even these had not been prepared in pure form. The commonest form of the element, which in itself is very far from common, is what is known to chemistry as chloride of radium which is a combination of chlorin and radium. This is a grayish white powder, somewhat like ordinary coarse table salt. To get enough to weigh a single grain requires the treatment of 1,200 pounds of pitchblende.

The second form of radium is as a bromide. In this form it costs $5,000 a grain and could a pound be obtained its value would be three-and-a-half million dollars.

Radium, as we understand it in any of its compounds, can communicate its property of radio-activity to other bodies. Any material when placed near radium becomes radio-active and retains such activity for a considerable time after being removed. Even the human body takes on this excited activity and this sometimes leads to annoyances as in delicate experiments the results may be nullified by the element acting upon the experimenter's person.

Despite the enormous amount of energy given off by radium it seems not to change in itself, there is no appreciable loss in weight nor apparently any microscopic or chemical change in the original body. Professor Becquerel has stated that if a square centimeter of surface was covered by chemically pure radium it would lose but one thousandth of a milligram in weight in a million years' time.

Radium is a body which gives out energy continuously and spontaneously. This liberation of energy is manifested in the different effects of its radiation and emanation, and especially in the development of heat. Now, according to the most fundamental principles of modern science, the universe contains a certain definite provision of energy which can appear under various forms, but which cannot be increased. According to Sir Oliver Lodge every cubic millimeter of ether contains as much energy as would be developed by a million horse power station working continuously far forty thousand years. This assertion is probably based on the fact that every corpuscle in the ether vibrates with the speed of light or about 186,000 miles a second.

It was formerly believed that the atom was the smallest sub-division in nature. Scientists held to the atomic theory for a long time, but at last it has been exploded, and instead of the atom being primary and indivisible we find it a very complex affair, a kind of miniature solar system, the centre of a varied attraction of molecules, corpuscles and electrons. Had we held to the atomic theory and denied smaller sub-divisions of matter there would be no accounting for the emissions of radium, for as science now believes these emissions are merely the expulsion of millions of electrons.

Radium gives off three distinct types of rays named after the first three letters of the Greek alphabet -

Alpha, Beta, Gamma - besides a gas emanation as does thorium which is a powerfully radio-active substance. The Alpha rays constitute ninety-nine per cent, of all the rays and consist of positively electrified particles. Under the influence of magnetism they can be deflected. They have little penetrative power and are readily absorbed in passing through a sheet of paper or through a few inches of air.

The Beta rays consist of negatively charged particles or corpuscles approximately one thousandth the size of those constituting the Alpha rays. They resemble cathode rays produced by an electrical discharge inside of a highly exhausted vacuum tube but work at a much higher velocity; they can be readily deflected by a magnet, they discharge electrified bodies, affect photographic plates, stimulate strongly phosphorescent bodies and are of high penetrative power.

The radiations are a million times more powerful than those of uranium. They have many curious properties.

If a photographic plate is placed in the vicinity of radium it is almost instantly affected if no screen intercepts the rays; with a screen the action is slower, but it still takes place even through thick folds, therefore, radiographs can be taken and in this way it is being utilized by surgery to view the anatomy, the internal organs, and locate bullets and other foreign substances in the system.

A glass vessel containing radium spontaneously charges itself with electricity. If the glass has a weak spot, a scratch say, an electric spark is produced at that point and the vessel crumbles, just like a Leyden jar when overcharged.

Radium liberates heat spontaneously and continuously. A solid salt of radium develops such an amount of heat that to every single gram there is an emission of one hundred calories per hour, in other words, radium can melt its weight in ice in the time of one hour.

As a result of its emission of heat radium has always a temperature higher by several degrees than its surroundings.

When a solution of a radium salt is placed in a closed vessel the radio-activity in part leaves the solution and distributes itself through the vessel, the sides of which become radio-active and luminous.

Radium acts upon the chemical constituents of glass, porcelain and paper, giving them a violet tinge, changes white phosphorous into yellow, oxygen into ozone and produces many other curious chemical changes.

We have said that it can serve the surgeon in physical examinations of the body after the manner of X-rays. It has not, however, been much employed in this direction owing to its scarcity and prohibitive price. It has given excellent results in the treatment of certain skin diseases, in cancer, etc. However it can have very baneful effects on animal organisms. It has produced paralysis and death in dogs, cats, rabbits, rats, guinea-pigs and other animals, and undoubtedly it might affect human beings in a similar way. Professor Curie said that a single gram of chemically pure radium would be sufficient to destroy the life of every man, woman and child in Paris providing they were separately and properly exposed to its influence.

Radium destroys the germinative power of seeds and retards the growth of certain forms of life, such as larvae, so that they do not pass into the chrysalis and insect stages of development, but remain in the state of larvae.

At a certain distance it causes the hair of mice to fall out, but on the contrary at the same distance it increases the hair or fur on rabbits.

It often produces severe burns on the hands and other portions of the body too long exposed to its activity.

It can penetrate through gases, liquids and all ordinary solids, even through many inches of the hardest steel. On a comparatively short exposure it has been known to partially paralyze an electric charged bar.

Heat nor cold do not affect its radioactivity in the least. It gives off but little light, its luminosity being largely due to the stimulation of the impurities in the radium by the powerful but invisible radium rays.

Radium stimulates powerfully various mineral and chemical substances near which it is placed. It is an infallible test of the genuineness of the diamond. The genuine diamond phosphoresces strongly when brought into juxtaposition, but the paste or imitation one glows not at all.

It is seen that the study of the properties of radium is of great interest. This is true also of the two other elements found in the ores of uranium and thorium, viz., polonium and actinium. Polonium, so-called, in honor of the native land of Mme. Curie, is just as active as radium when first extracted from the pitchblende but its energy soon lessens and finally it becomes inert, hence there has been little experimenting or investigation. The same may be said of actinium.

The process of obtaining radium from pitchblende is most tedious and laborious and requires much patience. The residue of the pitchblende from which uranium has been extracted by fusion with sodium carbonate and solution in dilute sulphuric acid, contains the radium along with other metals, and is boiled with concentrated sodium carbonate solution, and the solution of the residue in hydrochloric acid precipitated with sulphuric acid. The insoluble barium and radium sulphates, after being converted into chlorides or bromides, are separated by repeated fractional crystallization.

One kilogram of impure radium bromide is obtained from a ton of pitchblende residue after processes continued for about three months during which time, five tons of chemicals and fifty tons of rinsing water are used.

As has been said the element has never been isolated or separated in its metallic or pure state and most of the compounds are impure. Radium banks have been established in London, Paris and New York.

Whenever radium is employed in surgery for an operation about fifty milligrams are required at least and the banks let out the amount for about $200 a day. If purchased the price for this amount would be $4,000.

CHAPTER IV

MOVING PICTURES

Photographing Motion - Edison's Kinetoscope - Lumiere's Cinematographe - Before the Camera - The Mission of the Moving Picture.

Few can realize the extent of the field covered by moving pictures. In the dual capacity of entertainment and instruction there is not a rival in sight. As an instructor, science is daily widening the sphere of the motion picture for the purpose of illustration. Films are rapidly superseding text books in many branches. Every department capable of photographic demonstration is being covered by moving pictures. Negatives are now being made of the most intricate surgical operations and these are teaching the students better than the witnessing of the real operations, for at the critical moment of the operation the picture machine can be stopped to let the student view over again the way it is accomplished, whereas at the operating table the surgeon must go on with his work to try to save life and cannot explain every step in the process of the operation. There is no doubt that the moving picture machine will perform a very important part in the future teaching of surgery.

In the naturalist's domain of science it is already

playing a very important part. A device for microphotography has now been perfected in connection with motion machines whereby things are magnified to a great degree. By this means the analysis of a substance can be better illustrated than any way else. For instance a drop of water looks like a veritable Zoo with terrible looking creatures wiggling and wriggling through it, and makes one feel as if he never wanted to drink water again.

The moving picture in its general phase is entertainment and instruction rolled into one and as such it has superseded the theatre. It is estimated that at the present time in America there are upwards of 20,000 moving picture shows patronized daily by almost ten million people. It is doubtful if the theatre attendance at the best day of the winter season reaches five millions.

The moving picture in importance is far beyond the puny functions of comedy and tragedy. The grotesque farce of vaudeville and the tawdry show which only appeals to sentiment at highest and often to the base passions at lowest.

Despite prurient opposition it is making rapid headway. It is entering very largely into the instructive and the entertaining departments of the world's curriculum. Millions of dollars are annually expended in the production of films. Companies of trained and practiced actors are brought together to enact pantomimes which will concentrate within the space of a few minutes the most entertaining and instructive incidents of history and the leading happenings of the world.

At all great events, no matter where transpiring, the different moving picture companies have trained men at the front ready with their cameras to "catch" every incident, every movement even to the wink of an eyelash, so that the "stay-at-homes" can see the *show* as well, and with a great deal more comfort than if they had traveled hundreds, or even thousands, of miles to be present in *propria persona*.

How did moving pictures originate? What and when were the beginning? It is popularly believed that animated pictures had their inception with Edison who projected the biograph in 1887, having based it on that wonderful and ingenious toy, the Zoetrope. Long before 1887, however, several men of inventive faculties had turned their attention to a means of giving apparent animation to pictures. The first that met with any degree of success was Edward Muybridge, a photographer of San Francisco. This was in 1878. A revolution had been brought about in photography by the introduction of the instantaneous process. By the use of sensitive films of gelatine bromide of silver emulsion the time required for the action of ordinary daylight in producing a photograph had been reduced to a very small fraction of a second. Muybridge utilized these films for the photographic analysis of animal motion. Beside a race-track he placed a battery of cameras, each camera being provided with a spring shutter which was controlled by a thread stretched across the track. A running horse broke each thread the moment he passed in front of the camera and thus twenty or thirty pictures of him were taken in close succession within one or two seconds of time. From the negatives secured in this way a series of positives were obtained in proper order on a strip of sensitized paper. The strip when examined by means of the

Zoetrope furnished a reproduction of the horse's movements.

The Zoetrope was a toy familiar to children; it was sometimes called the wheel of life. It was a contrivance consisting of a cylinder some ten inches wide, open at the top, around the lower and interior rim of which a series of related pictures were placed. The cylinder was then rapidly rotated and the spectator looking through the vertical narrow slits on its outer surface, could fancy that the pictures inside were moving.

Muybridge devised an instrument which he called a Zoopraxiscope for the optical projection of his zoetrope photographs. The succession of positives was arranged in proper order upon a glass disk about 18 inches in diameter near its circumference. This disk was mounted conveniently for rapid revolution so that each picture would pass in front of the condenser of an optical lantern. The difficulties involved in the preparation of the disk pictures and in the manipulation of the zoopraxiscope prevented the instrument from attracting much attention. However, artistically speaking, it was the forerunner of the numerous "graphs" and "scopes" and moving picture machines of the present day.

It was in 1887 that Edison conceived an idea of associating with his phonograph, which had then achieved a marked success, an instrument which would reproduce to the eye the effect of motion by means of a swift and graded succession of pictures, so that the reproduction of articulate sounds as in the phonograph, would be accompanied by the reproduction of the motion naturally associated with them.

The principle of the instrument was suggested to Edison by the zoetrope, and of course, he well knew what Muybridge had accomplished in the line of motion pictures of animals almost ten years previously. Edison, however, did not employ a battery of cameras as Muybridge had done, but devised a special form of camera in which a long strip of sensitized film was moved rapidly behind a lens provided with a shutter, and so arranged as to alternately admit and cut off the light from the moving object. He adjusted the mechanism so that there were 46 exposures a second, the film remaining stationary during the momentary time of exposure, after which it was carried forward far enough to bring a new surface into the proper position. The time of the shifting was about one-tenth of that allowed for exposure, so that the actual time of exposure was about the one-fiftieth of a second. The film moved, reckoning shiftings and stoppages for exposures, at an average speed of a little more than a foot per second, so that a length of film of about fifty feet received between 700 and 800 impressions in a circuit of 40 seconds.

Edison named his first instrument the kinetoscope. It came out in 1893. It was hailed with delight at the time and for a short period was much in demand, but soon new devices came into the field and the kinetoscope was superseded by other machines bearing similar names with a like signification.

A variety of cameras was invented. One consisted of a film-feeding mechanism which moves the film step by step in the focus of a single lens, the duration of exposure being from twenty to twenty-five times as great as that necessary to move an unexposed portion of the film into position. No shutter was employed. As time passed many other improvements were made. An ingenious Frenchman named Lumiere, came forward with his Cinematographe which for a few years gave good satisfaction, producing very creditable results. Success, however, was due more to the picture ribbons than to the mechanism employed to feed them.

Of other moving pictures machines we have had the vitascope, vitagraph, magniscope, mutoscope, panoramagraph, theatograph and scores of others all derived from the two Greek roots *grapho* I write and *scopeo* I view.

The vitascope is the principal name now in use for moving picture machines. In all these instruments in order that the film projection may be visible to an audience it is necessary to have a very intense light. A source of such light is found in the electric focusing lamp. At or near the focal point of the projecting lantern condenser the film is made to travel across the field as in the kinetoscope. A water cell in front of the condenser absorbs most of the heat and transmits most of the light from the arc lamp, and the small picture thus highly illuminated is protected from injury. A projecting lens of rather short focus throws a large image of each picture on the screen, and the rapid succession of these completes the illusion of life-like motion.

Hundreds of patents have been made on cameras, projecting lenses and machines from the days of the kinetoscope to the present time when clear-cut moving pictures portray life so closely and so well as almost to deceive the eye. In fact in many cases the counterfeit is taken for the reality and audiences as much aroused as if they were looking upon a scene of actual life. We can well believe the story of the Irishman, who on seeing the stage villain abduct the young lady, made a rush at the canvas yelling out, - "Let me at the blackguard and I'll murder him."

Though but fifteen years old the moving picture industry has sent out its branches into all civilized lands and is giving employment to an army of thousands. It would be hard to tell how many mimic actors and actresses make a living by posing for the camera; their name is legion. Among them are many professionals who receive as good a salary as on the stage.

Some of the large concerns both in Europe and America at times employ from one hundred to two hundred hands and even more to illustrate some of the productions. They send their photographers and actors all over the world for settings. Most of the business, however, is done near home. With trapping and other paraphernalia a stage setting can be effected to simulate almost any scene.

Almost anything under the sun can be enacted in a moving picture studio, from the drowning of a cat to the hanging of a man; a horse race or fire alarm is not outside the possible and the aviator has been depicted "flying" high in the heavens.

The places where the pictures are prepared must be adapted for the purpose. They are called studios and have glass roofs and in most of them a good section of the walls are also glass. The floor space is divided into sections for the setting or staging of different productions, therefore several representations can take place at the same time before the eyes of the cameras.

There are "properties" of all kinds from the ragged garments of the beggar to kingly ermine and queenly silks. Paste diamonds sparkle in necklaces, crowns and tiaras, seeming to rival the scintillations of the Kohinoor.

At the first, objections were made to moving pictures on the ground that in many cases they had a tendency to cater to the lower instincts, that subjects were illustrated which were repugnant to the finer feelings and appealed to the gross and the sensual. Burglaries, murders and wild western scenes in which the villain-heroes triumphed were often shown and no doubt these had somewhat of a pernicious influence on susceptible youth. But all such pictures have for the most part been eliminated and there is a strict taboo on anything with a degrading influence or partaking of the brutal. Prize fights are often barred. In many large cities there is a board of censorship to which the different manufacturing firms must submit duplicates. This board has to pass on all the films before they are released and if the pictures are in any way contrary to morals or decency or are in any respect unfit to be displayed before the public, they cannot be put in circulation. Thus are the people protected and especially the youth who should be permitted to see nothing that is not elevating or not of a nature to inspire them with high and noble thoughts and with ambitions to make the world better and brighter.

Let us hope that the future mission of the moving picture will be along educational and moral lines tending to uplift and ennoble our boys and girls so that they may develop into a manhood and womanhood worthy the history and best traditions of our country.

* * * * * *

The Wizard of Menlo Park has just succeeded after two years of hard application to the experiment in giving us the talking picture, a real genuine talking picture, wholly independent of the old device of having the actors talk behind the screen when the films were projected. By a combination of the phonograph and the moving picture machine working in perfect synchronism the result is obtained. Wires are attached to the mechanism of both the machines, the one behind the screen and the one in front, in such a way that the two are operated simultaneously so that when a film is projected a corresponding record on the phonograph acts in perfect unison supplying the voice suitable to the moving action. Men and women pass along the canvas, act, talk, laugh, cry and "have their being" just as in real life. Of course, they are immaterial, merely the reflection of films, but the one hundred thousandth of an inch thick, yet they give forth oral sounds as creatures of flesh and blood. In fact every sound is produced harmoniously with the action on the screen. An iron ball is dropped and you hear its thud upon the floor, a plate is cracked and you can hear the cracking just the same as if the material plate were broken in your presence. An immaterial piano appears upon the screen and a fleshless performer discourses airs as real as those heard on Broadway. Melba and Tettrazini and Caruso and Bonci appear before you and warble their nightingale notes, as if behind the footlights with a galaxy of beauty, wealth and fashion before them for an audience. True it is not even their astral bodies you are looking at, only their pictured representations, but the magic of their voices is there all the same and there is such an atmosphere of realism about the representations that you can scarcely believe the actors are not present in *propriae personae*.

Mr. Edison had much study and labor of experiment in bringing his device to a successful issue. The greatest obstacle he had to overcome was in getting a phonograph that could "hear" far enough. At the beginning of the experiments the actor had to talk directly into the horn, which made the right kind of pictures impossible to get. Bit by bit, however, a machine was perfected which could "hear" so well that the actor could move at his pleasure within a radius of twenty feet. That is the machine that is being used now. This new combination of the moving picture machine and the phonograph Edison has named the *kinetophone*. By it he has made possible the bringing of grand opera into the hamlets of the West, and through it also our leading statesmen may address audiences on the mining camps and the wilds of the prairies where their feet have never trodden.

CHAPTER V

SKY-SCRAPERS AND HOW THEY ARE BUILT

Evolution of the Sky-scraper - Construction - New York's Giant Buildings - Dimensions.

The sky-scraper is an architectural triumph, but at the same time it is very much of a commercial enterprise, and it is indigenous, native-born to American soil. It had its inception here, particularly in New York and Chicago. The tallest buildings in the world are in New York. The most notable of these, the Metropolitan Life Insurance Building with fifty stories towering up to a height of seven hundred feet and three inches, has been the crowning achievement of architectural art, the highest building yet erected by man.

How is it possible to erect such building - how is it possible to erect a sky-scraper at all? A partial answer may be given in one word - *steel*.

Generally speaking the method of building all these huge structures is much the same. Massive piers or pillars are erected, inside which are usually strong steel columns; crosswise from column to column great girders are placed forming a base for the floor, and then upon the first pillars are raised other steel columns slightly decreased in size, upon which girders are again fixed for the next floor; and so on this process is continued floor after floor. There seems no reason why buildings should not be reared like this for even a hundred stories, provided the foundations are laid deep enough and broad enough.

The walls are not really the support of the buildings. The essential elements are the columns and girders of steel forming the skeleton framework of the whole. The masonry may assist, but the piers and girders carry the principal weight. If, therefore, everything depends upon these piers, which are often of steel and masonry combined, the immense importance will be seen of basing them upon adequate foundations. And thus it comes about that to build high we must dig deep, which fact may be construed as an aphorism to fit more subjects than the building of sky-scrapers.

To attempt to build a sky-scraper without a suitable foundation would be tantamount to endeavoring to build a house on a marsh without draining the marsh, - it would count failure at the very beginning. The formation depends on the height, the calculated weight the frame work will carry, the amount of air pressure, the vibrations from the running of internal machines and several other details of less importance than those mentioned, but of deep consequence in the aggregate.

Instead of being carried on thick walls spread over a considerable area of ground, the sky-scrapers are carried wholly on steel columns. This concentrates many hundred tons of load and develops pressure which would crush the masonry and cause the structures to penetrate soft earth almost as a stone sinks in water.

In the first place the weight of the proposed building and contents is estimated, then the character of the soil determined to a depth of one hundred feet if necessary. In New York the soil is treacherous and difficult, there are underground rivers in places and large deposits of sand so that to get down to rock bottom or pan is often a very hard undertaking.

Generally speaking the excavations are made to about a depth of thirty feet. A layer of concrete a foot or two thick is spread over the bottom of the pit and on it are bedded rows of steel beams set close together. Across the middle of these beams deep steel girders are placed on which the columns are erected. The heavy weight is thus spread out by the beams, girders and concrete so as to cause a reduced uniform pressure on the soil. Cement is filled in between the beams and girders and packed around them to seal them thoroughly against moisture; then clean earth or sand is rammed in up to the column bases and covered with the concrete of the cellar floor.

In some cases the foundation loads are so numerous that nothing short of masonry piers on solid rock will safely sustain them. To accomplish this very strong airtight steel or wooden boxes with flat tops and no bottoms are set on the pier sites at ground water level and pumped full of compressed air while men enter them and excavating the soil, undermine them, so they sink, until they land on the rock and are filled solid with concrete to form the bases of the foundation piers.

On the average the formation should have a resisting power of two tons to the square foot, dead load. By dead load is meant the weight of the steelwork, floors and walls, as distinguished from the office furniture and occupants which come under the head of living load. Some engineers take into consideration the pressure of both dead and live loads gauging the strength of the foundation, but the dead load pressure of 2 tons to the square foot will do for the reckoning, for as a live load only exerts a pressure of 60 lbs. to the square foot it may be included in the former.

The columns carry the entire weights including dead and live loads and the wind pressure, into the footings, these again distributing the loads on the soil. The aim is to have an equal pressure per square foot of soil at the same time, for all footings, thus insuring an even settlement. The skeleton construction now almost wholly consists of wrought steel. At first cast-iron and wrought-iron were used but it was found they corroded too quickly.

There are two classes of steel construction, the cage and the skeleton. In the cage construction the frame is strengthened for wind stresses and the walls act as curtains. In the skeleton, the frame carries only the vertical loads and depends upon the walls for its wind bracing. It has been found that the wind pressure is about 30 lbs. for every square foot of exposed surface.

The steel columns reach from the foundation to the top, riveted together by plates and may be extended to an indefinite height. In fact there is no engineering limit to the height.

The outside walls of the sky-scraper vary in thickness with the height of the building and also vary in accordance with the particular kind of construction, whether cage or skeleton. If of the cage variety, the walls, as has been said, act as curtains and consequently they are thinner than in the skeleton type of construction. In the latter case the walls have to resist the wind pressure unsupported by the steel frame and therefore they must be of a sufficient width. Brick and terra-cotta blocks are used for construction generally.

Terra-cotta blocks are also much used in the flooring, and for this purpose have several advantages over other materials; they are absolutely fire-proof, they weigh less per cubic foot than any other kind of fire-proof flooring and they are almost sound-proof. They do equally well for flat and arched floors.

It is of the utmost importance that the sky-scraper be absolutely fire-proof from bottom to top. These great buzzing hives of industry house at one time several thousand human beings and a panic would entail a fearful calamity, and, moreover, their height places the upper stories beyond reach of a water-tower and the pumping engines of the street.

The sky-scrapers of to-day are as fireproof as human ingenuity and skill can make them, and this is saying much; in fact, it means that they cannot burn. Of course fires can break out in rooms and apartments in the manufacturing of chemicals or testing experiments, etc., but these are easily confined to narrow limits and readily extinguished with the apparatus at hand. Steel columns will not burn, but if exposed to heat of sufficient degree they will warp and bend and probably collapse, therefore they should be protected by heat resisting agents. Nothing can be better than terra-cotta and concrete for this purpose. When terra-cotta blocks are used they should be at least 2 inches thick with an air space running through them. Columns are also fire-proofed by wrapping expanded metal or other metal lathing around them and plastering.

Then a furring system is put on and another layer of metal, lathing and plastering. This if well done is probably safer than the layer of hollow tile.

The floor beams should be entirely covered with terra-cotta blocks or concrete, so that no part of them is left exposed. As most office trimmings are of wood care should be taken that all electric wires are well insulated. Faulty installation of dynamos, motors and other apparatus is frequently the cause of office fires.

The lighting of a sky-scraper is a most elaborate arrangement. Some of them use as many lights as would well supply a good sized town. The Singer Building in New York has 15,000 incandescent lamps and it is safe to say the Metropolitan Life Insurance Building has more than twice this number as the floor area of the latter is 2-1/2 times as great. The engines and dynamos are in the basement and so fixed that their vibrations do not affect the building. As space is always limited in the basements of sky-scrapers direct connected engines and dynamos are generally installed instead of belt connected and the boilers operated under a high steam pressure. Besides delivering steam to the engines the boilers also supply it to a variety of auxiliary pumps, as boiler-feed, fire-pump, blow-off, tank-pump and pump for forcing water through the building.

The heating arrangement of such a vast area as is covered by the floor space of a sky-scraper has been a very difficult problem but it has been solved so that the occupant of the twentieth story can receive an equal degree of heat with the one on the ground floor. Both hot water and steam are utilized. Hot water heating, however, is preferable to steam, as it gives a much steadier heat. The radiators arc proportioned to give an average temperature of 65 degrees F. in each room during the winter months. There are automatic regulating devices attached to the radiators, so if the temperature rises above or falls below a certain point the steam or hot water is automatically turned on or off. Some buildings are heated by the exhaust steam from the engines but most have boilers solely for the purpose.

The sanitary system is another important feature. The supplying of water for wash-stands, the dispositions of wastes and the flushing of lavatories tax all the skill of the mechanical engineer. Several of these mighty buildings call for upwards of a thousand lavatories.

In considering the sky-scraper we should not forget the role played by the electric elevator. Without it these buildings would be practically useless, as far as the upper stories are concerned. The labor of stair climbing would leave them untenanted. No one would be willing to climb ten, twenty or thirty flights and tackle a day's work after the exertion of doing so. To climb to the fiftieth story in such a manner would be well-nigh impossible or only possible by relays, and after one would arrive at the top he would be so physically exhausted that both mental and manual endeavor would be out of the question. Therefore the elevator is as necessary to the skyscraper as are doors and windows. Indeed were it not for the introduction of the elevator the business sections of our large cities would still consist of the five and six story structures of our father's time instead of the towering edifices which now lift their heads among the clouds.

Regarded less than half a century ago as an unnecessary luxury the elevator to-day is an imperative necessity. Sky-scrapers are equipped with both express and local elevators. The express elevators do not stop until about the tenth floor is reached. They run at a speed of about ten feet per second. There are two types of elevators in general use, one lifting the car by cables from the top, and the other with a hydraulic plunger acting directly upon the bottom of the car. The former are operated either by electric motors or hydraulic cylinders and the latter by hydraulic rams, the cylinders extending the full height of the building into the ground.

America is pre-eminently the land of the sky-scraper, but England and France to a degree are following along the same lines, though nothing as yet has been erected on the other side of the water to equal the towering triumphs of architectural art on this side. In no country in the world is space at such a premium as in New York City, therefore, New York *per se* may be regarded as the true home of the tall building, although Chicago is not very much behind the Metropolis in this respect.

As figures are more eloquent than words in description the following data of the two giant structures of the Western World may be interesting.

The Singer Building at the corner of Broadway and Liberty Street, New York City, has a total height from the basement floor to the top of the flagstaff of 742 feet; the height from street to roof is 612 feet, 1 inch. There are 41 stories. The weight of the steel in the entire building is 9,200 tons. It has 16 elevators, 5 steam engines, 5 dynamos, 5 boilers and 28 steam pumps. The length of the steam and water piping is 5 miles. The cubical contents of the building comprise 66,950,000 cubic feet, there are 411,000 square feet of floor area or about 9-1/2 acres. The weight of the tower is 18,300 tons. Little danger from a collapse will be apprehended when it is learned that the columns are securely bolted and caissons which have been sunk to rock-bed 80 feet below the curb.

The other campanile which has excited the wonder and admiration of the world is the colossal pile known as the Metropolitan Building. This occupies the entire square or block as we call it from 23rd St. to 24th St. and from Madison to Fourth Avenue. It is 700 feet and 3 inches above the sidewalk and has 50 stories. The main building which has a frontage of 200 feet by 425 feet is ten stories in height. It is built in the early Italian renaissance style the materials being steel and marble. The Campanile is carried up in the same style and is also of marble. It stands on a base measuring 75 by 83 feet and the architectural treatment is chaste, though severe, but eminently agreeable to the stupendous proportions of the structure. The tower is quite different from that of the Singer Building. It has twelve wall and eight interior columns connected at every fourth floor by diagonal braces; these columns carry 1,800 pounds to the linear foot. The wind pressure calculated at the rate of 30 lbs. to the square foot is enormous and is provided for by deep wall girders and knee braces which transfer the strain to the columns and to the foundation. The average cross section of the tower is 75 by 85 feet, the floor space of the entire building is 1,080,000 square feet or about 25 acres.

The tower of this surpassing cloud-piercing structure can be seen for many miles from the surrounding country and from the bay it looks like a giant sentinel in white watching the mighty city at its feet and proclaiming the ceaseless activity and progress of the Western World.

CHAPTER VI

OCEAN PALACES

Ocean Greyhounds - Present Day Floating Palaces - Regal Appointments - Passenger Accommodation - Food Consumption - The One Thousand Foot Boat.

The strides of naval architecture and marine engineering have been marvelous within the present generation. To-day huge leviathans glide over the waves with a swiftness and safety deemed absolutely impossible fifty years ago.

In view of the luxurious accommodations and princely surroundings to be found on the modern ocean palaces, it is interesting to look back now almost a hundred years to the time when the *Savannah* was the first steamship to cross the Atlantic. True the voyage of this pioneer of steam from Savannah to Liverpool was not much of a success, but she managed to crawl across the sails very materially aiding the engines, and heralded the dawn of a new day in transatlantic travel. No other steamboat attempted the trip for almost twenty years after, until in 1838 the *Great Western* made the run in fifteen days. This revolutionized water travel and set the whole world talking. It was the beginning of the passing of the sailing ship and was an event for rejoicing. In the old wooden hulks with their lazily flapping wings, waiting for a breeze to stir them, men and women and children huddled together like so many animals in a pen, had to spend weeks and months on the voyage between Europe and America. There was little or no room for sanitation, the space was crowded, deadly germs lurked in every cranny and crevice, and consequently hundreds died. To many indeed the sailing ship became a floating hearse.

In those times, and they are not so remote, a voyage was dreaded as a calamity. Only necessity compelled the undertaking. It was not travel for pleasure, for pleasure under such circumstances and amid such surroundings was impossible. The poor emigrants who were compelled through stress and poverty to leave their homes for a foreign country feared not toil in a new land, but they feared the long voyage with its attending horrors and dangers. Dangerous it was, for most of the sailing vessels were unseaworthy and when a storm swept the waters, they were as children's toys, at the mercy of wind and wave. When the passenger stepped on board he always had the dread of a watery grave before him.

How different to-day. Danger has been eliminated almost to the vanishing point and the mighty monsters of steel and oak now cut through the waves in storms and hurricanes with as much ease as a duck swims through a pond.

From the time the *Great Western* was launched, steamships sailing between American and English ports became an established institution. Soon after the *Great Western's* first voyage a sturdy New England Quaker from Nova Scotia named Samuel Cunard went over to London to try and interest the British government in a plan to establish a line of steamships between the two countries. He succeeded in raising 270,000 pounds, and built the *Britannia*, the first Cunard vessel to cross the Atlantic. This was in 1840. As ships go now she was a small craft indeed. Her gross tonnage was 1,154 and her horse power 750. She carried only first-class passengers and these only to the limit of one hundred. There was not much in the way of accommodation as the quarters were cramped, the staterooms small and the sanitation and ventilation defective. It was on the *Britannia* that Charles Dickens crossed over to America in 1842 and he has given us in his usual style a pen picture of his impressions aboard. He stated that the saloon reminded him of nothing so much as of a hearse, in which a number of half-starved stewards attempted to warm themselves by a glimmering stove, and that the staterooms so-called were boxes in which the bunks were shelves spread with patches of filthy bed-clothing, somewhat after the style of a mustard plaster. This criticism must be taken with a little reservation. Dickens was a pessimist and always censorious and as he had been feted and feasted with the fat of the land, he expected that he should have been entertained in kingly quarters on shipboard. But because things did not come up to his expectations he dipped his pen in vitriol and began to criticise.

At any rate the *Britannia* in her day was looked upon as the *ne plus ultra* in naval architecture, the very acme of marine engineering. The highest speed she developed was eight and one-half knots or about nine and three-quarters miles an hour. She covered the passage from Liverpool to Boston in fourteen and one-half days, which was then regarded as a marvellous feat and one which was proclaimed throughout

England with triumph.

For a long time the *Britannia* remained Queen of the Seas for speed, but in 1852 the Atlantic record was reduced to nine and a half days by the *Arctic*. In 1876 the *City of Paris* cut down the time to eight days and four hours. Twelve years later in 1879 the *Arizona* still further reduced it to seven days and eight hours. In 1881 the *Alaska*, the first vessel to receive the title of "*Ocean Greyhound*," made the trip in six days and twenty-one hours; in 1885 the *Umbria* bounded over in six days and two hours, in 1890 the *Teutonic* of the White Star line came across in five days, eighteen hours and twenty-eight minutes, which was considered the limit for many years to come. It was not long however, until the Cunard lowered the colors of the White Star, when the *Lucania* in 1893 brought the record down to five days and twelve hours. For a dozen years or so the limit of speed hovered round the five-and-a-half day mark, the laurels being shared alternately by the vessels of the Cunard and White Star Companies. Then the Germans entered the field of competition with steamers of from 14,500 to 20,000 tons register and from 28,000 to 40,000 horse power. The *Deutschland* soon began setting the pace for the ocean greyhounds, while other vessels of the North German Lloyd line that won transatlantic honors were the *Kaiser Wilhelm II., Kaiser Wilhelm der Grosse, Kronprinz Wilhelm and Kronprinzessin Cecilie*, all remarkably fast boats with every modern luxury aboard that science could devise. These vessels are equipped with wireless telegraphy, submarine signalling systems, water-tight compartments and every other safety appliance known to marine skill. The *Kaiser Wilhelm der Grosse* raised the standard of German supremacy in 1902 by making the passage from Cherbourg to Sandy Hook lightship in five days and fifteen hours.

In 1909, however, the sister steamships *Mauretania* and *Lusitania* of the Cunard line lowered all previous ocean records, by making the trip in a little over four and a half days. They have been keeping up this speed to the present time, and are universally regarded as the fastest and best equipped steamships in the world, - the very last word in ocean travel. On her last mid-September voyage the *Mauretania* has broken all ocean records by making the passage from Queenstown to New York in 4 days 10 hours and 47 minutes. But they are closely pursued by the White Star greyhounds such as the *Oceanic*, the *Celtic* and the *Cedric*, steamships of world wide fame for service, appointments, and equipment. Yet at the present writing the Cunard Company has another vessel on the stocks, to be named the *Falconia* which in measurements will eclipse the other two and which they are confident will make the Atlantic trip inside four days.

The White Star Company is also building two immense boats to be named the *Olympic* and *Titanic*. They will be 840 feet in length and will be the largest ships afloat. However, it is said that freight and passenger-room is being more considered in the construction than speed and that they will aim to lower no records. Each will be able to accommodate 5,000 passengers besides a crew of 600.

All the great liners of the present day may justly be styled ocean palaces, as far as luxuries and general appointments are concerned, but as the *Mauretania* and *Lusitania* are best known, a description of either of these will convey an idea to stay-at-homes of the regal magnificence and splendors of the floating hotels which modern science places at the disposal of the traveling public.

Though sister ships and modeled on similar lines, the *Mauretania* and *Lusitania* differ somewhat in construction. Of the two the *Mauretania* is the more typical ship as well as the more popular. This modern triumph of the naval architect and marine engineer was built by the firm of Swan, Hunter & Co. at Wellsend on the Tyne in 1907. The following are her dimensions: Length over all 790 feet. Length between perpendiculars 760 feet. Breadth 88 feet. Depth, moulded 60.5 feet. Gross tonnage 32,000. Draught 33.5 feet. Displacement 38,000 tons.

She has accommodation space for 563 first cabin, 500 second cabin, and 1,300 third class passengers. She carries a crew of 390 engineers, 70 sailors, 350 stewards, a couple of score of stewardesses, 50 cooks, the officers and captain, besides a maritime band, a dozen or so telephone and wireless telegraph operators, editor and printers for the wireless bulletin published on board and two attendants for the elevator.

The type of engine is what is known as the Parsons Turbine. There are 23 double ended and 2 single ended boilers. The engines develop 68,000 horse power; they are fed by 192 furnaces; the heating surface is 159,000 square feet; the grate surface is 4,060 square feet; the steam pressure is 195 lbs. to the square inch.

The highest speed attained has been almost 26 knots or 30 miles an hour. At this rate the number of revolutions is 180 to the minute. The coal daily consumed by the fiery maw of the furnaces is enormous. On one trip between Liverpool and New York more than 7,000 tons is required which is a consumption of over 1,500 tons daily.

There are nine decks, seven of which are above the water line. Corticine has been largely used for deck covering, instead of wood as it is much lighter. On the boat deck which extends over the greater part of the centre of the ship are located several of the beautiful *en suite* cabins. Abaft these at the forward end are the grand Entrance Hall, the Library, the Music-Room and the Lounging-Room and Smoking-Room for the first cabin passengers.

There is splendid promenading space on the boat deck where passengers can exercise to their hearts' content and also indulge in games and sports with all the freedom of field life. Many life boats swing on davits and instead of being a hindrance or obstacle, act as shades from the sunshine and as breaks from the wind.

In the space for first-class passengers are arranged a large number of cabins. What are known as the regal suites are on both port and starboard, and along each side of the main deck are more *en suite* rooms.

On the shelter deck there are no first-class cabin quarters. At the forward end of this deck are the very powerful Napier engines for working the anchor gear. Abaft this on the starboard side is the general lounging room for third-class passengers, while on the port-side is their smoking room with a companion way leading to the third-class dining saloon below and to the third-class cabins on the main and lower decks. The third-class galleys are accommodated on the main deck house and close by is a set of the refrigerating machinery used in connection with the rooms for the storage of supplies for the kitchen department. The side of the ship for a considerable distance aft of this is plated up to the promenade deck level so that the third-class passengers have not only convenient rooms but a protected promenade. Abaft this promenade is another open one. Indeed the accommodations for the third class are as good as what the first-class were accustomed to on most of the liners some dozen years ago.

To the left of the grand staircase on the deck house is a children's dining saloon and nursery.

On the top deck are dining saloons for all three classes of passengers, that for the third being forward, for the first amidships and for the second near the stern; 470 first-class passengers can be seated at a time, 250 second class and more than 500 of the third class.

The main deck is given up entirely to staterooms. The whole of the lower deck forward is also arranged for third-class staterooms. The firemen and other engine room and stokehold workers are located in rooms above the machinery with separate entrances and exits to and from their work. Promenade and exercise space is provided for them on the shelter deck which is fenced off from the space of the second and third class passenger. Amidships is a coal bunker with a compartment under the engines for the storage of supplies.

The coal trimmers are accommodated alongside the engine casing and abaft this are the mailrooms with accommodation for the stewards and other helpers. The "orlop" or eighth deck is devoted entirely to machinery with coal bunkers on each side of the boilers to provide against the effect of collisions.

The general scheme of color throughout the ship is pleasing and harmonious. The wood for the most part is oak and mahogany. There are over 50,000 square feet of oak in parquet flooring. All the carving and tracing is done in the wood, no superpositions or stucco work whatever being used to show reliefs.

The grand stairway shows the Italian renaissance style of the 16th century; the panels are of French walnut; the carving of columns and pilasters is of various designs but the aggregate is pleasing in effect.

The Library extends across the deck house, 33 by 56 feet; the walls of the deck house are bowed out to form bay windows. When you first enter the Library the effect is as though you were looking at shimmering marble, this is owing to the lightness of the panels which are sycamore stained in light gray. The mantelpiece is of white statuary marble. The great swing doors which admit you, have bevelled glass panels set in bronze casings. The chairs have mahogany frames done in light plush.

The first class lounging room is probably the most artistic as well as the most sumptuous apartment in the ship. The panels are of beautiful ingrained mahogany dully polished a rich brown. The white ceiling is of simple design with boldly carved mouldings and is supported by columns embossed in gold of exquisite workmanship. Some of the panels are of curiously woven tapestries, the fruit of oriental looms. Chandeliers of beautiful design in rich bronze and crystal depend from the ceiling. The curtains, hanging with their soft folds against the dull gold of the carved curtainboxes, are of a charming cream silk and with their flower borders lend a tone both sumptuous and refined. The carpet is of a slender trellis design with bluish pink roses trailing over a pearl grey ground and forms a perfect foil to the splendid furniture. The chairs are of polished beech covered with 18th century brocade.

The smoking-room of the first-class is done in rich oak carving with an inlaid border around the panels. An unusual feature in the main part of the room is a jube passageway extending the whole length and divided into recesses with divans and card tables. Writing tables may be found in secluded nooks free from interruption. The windows of unusual size, are semi-circular and give a home-like appearance to the room.

The dining saloon is in light oak with all carvings worked in the wood. A children's nursery off the main stairway in the deck house is done in mahogany. Enameled white panels depict the old favorite of the Four and Twenty Blackbirds baked in a Pie.

An air of delicate refinement and rich luxury hangs about the regal rooms. A suite consists of drawing-room, dining-room, two bedrooms, bathroom and a private corridor. The drawing- and dining-rooms of these suites are paneled in East India satin-wood, probably the hardest and most durable of all timber. The bedrooms are in Georgian style finished in white with satin hangings.

The special staterooms are also finished in rich woods on white and gold and have damask and silk hangings and draperies. An idea of the richness and magnificence of the interior decorations may be obtained when it is learned that the cost of these decorations exceeded three million dollars.

The galleys, pantries, bakery, confectionery and utensil cleaning rooms extend the full length of the ship. Electricity plays an important part in the culinary department. Electric motors mix dough, run grills and roasters, clean knives and manipulate plate racks and other articles of the kitchen. The main cooking range for the saloon is 24 by 8 feet, heated by coal. There are four steam boilers and 12 steam ovens. There are extensive cold storage compartments and refrigerating chambers.

In connection with the commissariat department it is interesting to note the food supply carried for a trip of this floating caravansary. Here is a list of the leading supplies needed for a trip, but there are hundreds of others too numerous to mention: Forty thousand pounds of fresh beef, 1,000 lbs. of corned beef, 8,000 lbs. of mutton, 800 lbs. of lamb, 600 lbs. of veal, 500 lbs. of pork, 4,000 lbs. of fish, 2,000 fowls, 100 geese, 150 turkeys, 350 ducks, 400 pigeons, 250 partridges, 250 grouse, 200 pheasants, 800 quail, 200 snipe, 35 tons of potatoes, 75 hampers of vegetables, 500 quarts ice ream, 3,500 quarts of milk, 30,000 eggs and in addition many thousand bottles of mineral water and spirituous liquors.

The health of the passengers is carefully guarded during the voyage. The science of thermodynamics has been brought to as great perfection as possible. Not alone is the heating thoroughly up to modern science requirements but the ventilation as well, by means of thermo tanks, suction valves and exhaust fans. All foul air is expelled and fresh currents sent through all parts of the ship.

There is an electric generating station abaft the main

engine room containing four turbo-generators each of 375 kilowatts capacity.

There are more than 5,000 electric lights and every room is connected by an electric push-bell. There is a telephone exchange through which one can be connected with any department of the vessel. When in harbor, either at Liverpool or New York, the wires are connected to the City Central exchange so that the ships can be communicated with either by local or long distance telephone.

By means of wireless telegraphy voyagers can communicate with friends during almost the entire trip and learn the news of the world the same as if they were on land. A bulletin is published daily on board giving news of the leading happenings of the world.

There is a perfect fire alarm system on board with fire mains on each side of the ship from which connections are taken to every separate department. There are boxes with hydrant and valve in each room and a system of break glass fire alarms with a drop indicator box in the chartroom and also one in the engine-room to notify in case of any outbreak.

The sanitation is all that could be desired. There are flush lavatories on all decks in marble and onyx and with all the sanitary contrivances in apparatus of the best design.

The vessel is propelled by four screws, rotated by turbine engines and the power developed is equal to that of 68,000 horses. Now 68,000 horses placed head to tail in a single line would reach a distance of 90 miles or as far as from New York to Philadelphia; and if the steeds were harnessed twenty abreast there would be no fewer than 3,400 rows of powerful horses.

Such is the steamship of to-day but there is no doubt that the thousand foot boat is coming, which probably will cross the Atlantic ocean in less than four days if not in three. But the question is, where shall we put her, that is, where shall we dock her?

To build a thousand foot pier to accommodate her, appears like a good answer to this question, but the great difficulty is that there are United States Government regulations restricting the length of piers to 800 feet. Docking space along the shore of New York harbor is too valuable to permit the ship being berthed parallel to the shore, therefore vessels must dock at right angles to the shore. Some provisions must soon be made and the regulations as to dock lengths revised.

The thousand footer may be here in a couple of years or so. In the meantime the two 840 footers are already on the stocks at Belfast and are expected to arrive early in 1911. Before they come changes and improvements must be made in the docking and harbor facilities of the port of New York.

If higher speed is demanded, increased size is essential, since with even the best result every 100 horse-power added involves an addition to machinery weight of approximately 14 tons and to the area occupied of about 40 square feet. To accomplish this the ship must be as much larger in proportion.

The ship designer has to work within circumscribed limits. If he could make his vessel of any depth he might build much larger and there would be theoretically no limit to his speed: 40 knots an hour might be obtained as easily as the present maximum of 26, but in designing his ship he must remember that in the harbors of New York or Liverpool the channels are not much beyond 30 feet in depth. High speed necessitates powerful engines, but if the engines be too large there will not be space enough for coal to feed the furnaces. If the breadth of the ship is increased the speed is diminished, while on the other hand, if too powerful engines are put in a narrow vessel she will break her back. The proper proportions must be carefully studied as regards length, breadth, depth and weight so that the vessel will derive the greatest speed from her engines.

CHAPTER VII

WONDERFUL CREATIONS IN PLANT LIFE

Mating Plants - Experiments of Burbank - What he has Accomplished.

In California lives a wonderful man. He has succeeded in doing more than making two blades of grass grow where grew but one. Yearly, daily in fact, this wizard of plant life is playing tricks on old Mother Nature, transforming her vegetable children into different shapes and making them no longer recognizable in their original forms. Like the fairies in Irish mythology, this man steals away the plant babies, but instead of leaving sickly elves in their places, he brings into the world exceedingly healthy or lusty youngsters which grow up into a full maturity, and develop traits of character superior to the ones they supplant. For instance he took away the ugly, thorny insipid cactus and replaced it by a beautiful smooth juicy one which is now making the western deserts blossom as the rose. The name of this man is Luther Burbank whose fame as a creator of new plants has become world wide.

The basic principle of Burbank's plant magic comes under two heads, viz.: breeding and selection. He mates two different species in such a way that they will propagate a type partaking of the natures of both but superior to either in their qualities. In order to effect the best results from mating, he is choice in his selection of species - the best is taken and the worst rejected. It is a universal law that the bad can never produce the good; consequently when good is desired, as is universally the case, bad must be eliminated. In his method, Burbank gives the good a chance to assert itself and at the same time takes away all opportunity from the bad. So that the latter cannot thrive but must decay and pass out of being. He takes two plants - they may be of the same species, but as a general rule he prefers to experiment with those of different species; he perceives that neither one in its present surroundings is putting forth what is naturally expected from it, that each is either retrograding in the scale of life or standing still for lack of encouragement to go forward. He knows that back of these plants is a long history of evolutions from primitive beginnings to their present stage just as in the case of man himself. 'Tis a far cry from the cliff-dweller wielding his stone-axe and roaming nude through the fields and forests after his prey - the wild beast - to the lordly creature of to-day - the product of long ages of civilization and culture, yet high as the state is to which man has been brought, in many cases he is hemmed in and surrounded by circumstances which preclude him from putting forth the best that is in him and showing his full possibilities to the world. The philosopher is often hidden in the ploughman and many a poor laborer toiling in corduroys and fustian at the docks, in the mills, or sweeping the streets may have as good a brain as Edison, but has not the opportunity to develop it and show its capabilities. The same analogy is applicable to plant life. Under adverse conditions a plant or vegetable cannot put forth its best efforts. In a scrawny, impoverished soil, and exhausted atmosphere, lacking the constituents of nurture, the plant will become dwarfed and unproductive, whereas on good ground and in good air, which have the succulent properties to nourish it the best results may be expected. The soil and the air, therefore, from which are derived the constituents of plant life, are indispensably necessary, but they are not the primal principles upon which that life depends for its being. The basis, the foundation, the origin of the life is the seed which germinates in the soil and evolves itself into the plant.

A dead seed will not germinate, a contaminated seed may, but the plant it produces will not be a healthy one and it will only be after a long series of transplantings, with patience and care, that at length a really sound plant will be obtained. The same principle holds good in regard to the human plant. It is hard to offset an evil ancestry. The contamination goes on from generation to generation, just as in the case of the notorious Juke family which cost New York State hundreds of thousands of dollars in consequence of criminality and idiocy. It requires almost a miracle to divert an individual sprung from a corrupt stem into a healthy, moral course of living. There must be some powerful force brought to bear to make him break the ligatures which bind him to ancestral nature and enable him to come forth on a plane where he will be susceptible to the influence of what is good and noble. Such can be done and has been accomplished.

Burbank is accomplishing such miracles in the vegetable kingdom, in fact he is recreating species as it were and developing them to a full fruition. Of course as in the case of the conversion of a sinner from his evil instincts, much opposition is met and the progress at first is slow, but finally the plant becomes fixed in its new ways and starts forward on its new course in life. It requires patience to await the development Burbank is a man of infinite patience. He has been five, ten, fifteen, twenty years in producing a desired blossom, but he considers himself well rewarded when his object has been obtained. Thousands of experiments are going on at the same time, but in each case years are required to achieve results, so slow is the work of selection, the rejecting of the seemingly worthless and the eternal choosing of the best specimens to continue the experiments.

When two plants are united to produce a third, no human intelligence can predict just what will be the result of the union. There may be no result at all; hence it is that Burbank does not depend on one experiment at a time. If he did the labors of a life-time would have little to show for their work. In breeding lilies he has used as high as five hundred thousand plants in a single test. Such an immense quantity gave him a great variety of selection. He culled and rejected, and culled and rejected until he made his final selection for the last test.

Sometimes he is very much disappointed in his anticipations. For instance, he marks out a certain life for a flower and breeds and selects to that end. For a time all may go according to his plans, but suddenly some new trait develops which knocks those plans all out of gear. The new flower may have a longer stem and narrower leaves than either parent, while a shorter stem and broader leaves are the desideratum. The experimenter is disappointed, but not disheartened; he casts the flower aside and makes another selection from the same species and again goes ahead, until his object is attained.

It may be asked how two plants are united to procure a third. The act is based on the procreative law of nature. Plant-breeding is simply accomplished by sifting the pollen of one plant upon the stigma of another, this act - pollenation - resulting in fertilization, Nature in her own mysterious ways bringing forth the new plant.

In order to get an idea of the Burbank method, let us consider some of his most famous experiments, for instance, that in which by uniting the potato with the tomato he has produced a new variety which has been very aptly named the pomato. Mr. Burbank, from the beginning of his wonderful career, has experimented much with the potato. It was this vegetable which first brought the plant wizard into worldwide prominence. The Burbank potato is known in all lands where the tuber forms an article of food. It has been introduced into Ireland and promises to be the salvation of that distressed island of which the potato constitutes the staple diet. The Burbank potato is the hardiest of all varieties and in this respect is well suited for the colder climates of the Temperate Zone. Apart from this potato which bears his name, Mr. Burbank has produced many other varieties. He has blended wild varieties with tame ones, getting very satisfactory results. Mr. Burbank believes that a little wild blood, so to speak, is often necessary to give tone and vigor to the tame element which has been long running in the same channels. Probably it was Emerson, his favorite author, who gave him the cue for this idea. Emerson pointed out that the city is recruited from the country. "The city would have died out, rotted and exploded long ago," wrote the New England sage, "but that it was reinforced from the fields. It is only country that came to town day before yesterday, that is city and court to-day."

In Burbank's greenhouses are mated all kinds of wild and tame varieties of potatoes, producing crosses and combinations truly wonderful as regards shape, size, and color. One of the most palatable potatoes he has produced is a magenta color approaching crimson, so distributed throughout that when the tuber is cut, no matter from what angle, it presents concentric geometric figures, some having a resemblance to human and animal faces.

Before entering on any experiment to produce a new creation, Burbank always takes into consideration the practical end of the experiment, that is, what the value of the result will be as a practical factor in commerce, how much it will benefit the race. He does not experiment for a pastime or a novelty, but for a purpose. His object in regard to the potato is to make it a richer, better vegetable for a food supply and also to make it more important for other purposes in the commerce of the nations.

The average potato consists of seventy-five per cent. water and twenty-five per cent. dry matter, almost all of which is starch. Now starch is a very important article from a manufacturing standpoint, but only one-fourth of the potato is available for manufacturing, the other three-fourths, being water, is practically waste matter. Now if the water could be driven out to a great extent and starchy matter increased it is easy to understand that the potato would be much increased in value as an article of manufacture. Burbank has not overlooked this fact in his potato experiments. He has demonstrated that it is as easy to breed potatoes for a larger amount of starch, and he has really developed tubers which contain at least twenty-five per cent. more starch than the normal varieties; in other words, he has produced potatoes which yield fifty per cent. of starch instead of twenty-five per cent. The United States uses about $12,000,000 worth of starch every year, chiefly obtained from Indian corn and potatoes. When the potato is made to yield double the amount of starch, as Burbank has proved it can yield and more, it will be understood what a large part it can be made to play in this important manufacture.

Also for the production of alcohol the potato is gaining a prominent place. The potato starch is converted into maltose by the diastase of malt, the maltose being easily acted upon by ferment for the actual production of the alcohol. Therefore an increase in the starch of the potato for this purpose alone is much to be desired.

Of course the chief prominence of the potato will still consist in its adaptability as an article of food. Burbank does not overlook this. He has produced and is producing potatoes with better flavor, of larger and uniform size and which give a much greater yield to the area. Palatability in the end decides the permanence of a food, and the Burbank productions possess this quality in a high degree.

Burbank labored long and studied every characteristic of the potato before attempting any experiments with the tomato. Though closely related by family ties, the potato and the tomato seemed to have no affinity for each other whatever. In many other instances it has also been found that two varieties which from a certain relation might naturally be expected to amalgamate easily have been repellant to each other and refused to unite.

In his first experiment in trying to cross the potato and tomato, Burbank produced tomatoes from the seeds of plants pollenated from potato pollen only. He next produced what he called "aerial potatoes" of very peculiar twisted shapes from a potato vine grafted on a Ponderosa or large tomato plant. Then reversing this operation he grafted the same kind of tomato plant upon the same kind of potato plant and produced underground a strange-looking potato with marked tomato characteristics. He saw he was on the right road to the production of a new variety of vegetable, but before experimenting further along this line he crossed two distinct species of tomatoes and obtained a most ornamental plant, different from the parent stems, about twelve inches high and fifteen inches across with large unusual leaves and producing clusters of uniform globular fruit, the whole giving a most pleasing and unique appearance. The fruit were more palatable than the ordinary tomatoes, had better nutritive qualities and were more suitable for preserving and canning.

Very pleased with this result he went back to his experiments with the potato-tomato, and succeeded in producing the most wonderful and unique fruit in the world, one which by a happy combination of the two names, he has aptly called the pomato. It may be considered as the evolution of a potato seed-ball. It first appears as a tiny green ball on the potato top and as the season progresses it gradually enlarges and finally develops into a fruit about the size and shape of the ordinary tomato. The flesh is white and the marrow, which contains but a few tiny white seeds, is exceedingly pleasant to the taste, possessing a combination of several different fruit flavors, though it cannot be identified with any one. It may be eaten either raw or cooked after the manner of the common tomato. In either case it is most palatable, but especially so when cooked. It is exceptionally well adapted to preserving purposes.

The production of such a fruit from a vegetable is one of the crowning triumphs of the California wizard. Probably it is the most novel of all the wonderful crosses and combinations he has given to the world.

It would be impossible here to go into detail in regard to some of the other wonders accomplished in the plant world by this modern magician. There is only space to merely mention a few more of his successful achievements. He has given the improved thornless and spiculess cactus, food for man and beast, converting it into a beautifier and reclaimer of desert wastes; the plum-cot which is an amalgamation of the plum and the apricot with a flavor superior to both; many kinds of plums, some without pits, others having the taste of Bartlett pears, and still others giving out a fragrance as sweet as the rose; several varieties of walnuts, one with a shell as thin as paper and which was so easily broken by the birds that Burbank had to reverse his experiment somewhat in order to get a thicker shell; another walnut has no tannin in the meat, which is the cause of the disagreeable flavor of the ordinary fruit; the world-famed Shasta daisy, which is a combination of the Japanese daisy, the English daisy and the common field daisy, and which has a blossom seven inches in diameter; a dahlia deprived of its unpleasant odor and the scent of the magnolia blossom substituted; a gladiolus which blooms around the entire stem like a hyacinth instead of the old way on one side only; many kinds of lilies with chalices and petals different from the ordinary, and exhaling perfumes as varied as those of Oriental gardens; a poppy of such dimension that it is from ten to twelve inches across its brilliant bloom; an amaryllis bred up from a couple of inches to over a foot in diameter; several kinds of fruit trees which withstand frost in bud and in flower; a chestnut tree which bears nuts in eighteen months from the time of seed-planting; a white blackberry (paradoxical as it may appear), a rare and beautiful fruit and as palatable as it is beautiful; the primusberry, a union of the raspberry and the blackberry; another wonderful and delicious berry produced from the California dewberry and the Cuthbert-raspberry; pieplants four feet in diameter, bearing every day in the year; prunes, three, four, and five times as large as the ordinary and enriched in flavor; blackberries without their prickly thorns and hundreds of other combinations and crosses of fruits and flowers too numerous to mention. He has improved plums, pears, apples, apricots, quinces, peaches, cherries, grapes, in short, all kinds of fruit which grow in our latitude and many even that have been introduced. He has developed hundreds of varieties of flowers, improving them in color, hardiness and yield. Thus he has not only added to the food and manufacturing products of the world, but he has enriched the aesthetic side in his beautiful flower creations.

CHAPTER VIII

LATEST DISCOVERIES IN ARCHAEOLOGY

Prehistoric Time - Earliest Records - Discoveries in Bible Lands - American Explorations.

For the earliest civilization and culture we must go to that part of the world, which according to the general belief, is the cradle of the human race. The civilization of the Mesopotamian plain is not only the oldest but the first where man settled in great city communities, under an orderly government, with a developed religion, practicing agriculture, erecting dwellings and using a syllabified writing. All modern civilization had its source there. For 6,000 years the cuneiform or wedge-shaped writing of the Assyrians was the literary script of the whole civilized ancient world, from the shores of the Mediterranean to India and even to China, for Chinese civilization, old as it is, is based upon that which obtained in Mesopotamia. In Egypt, too, at an early date was a high form of neolithic civilization. Six thousand years before Christ, a white-skinned, blond-haired, blue-eyed race dwelt there, built towns, carried on commerce, made woven linen cloth, tanned leather, formed beautiful pottery without the wheel, cut stone with the lathe and designed ornaments from ivory and metals. These were succeeded by another great race which probably migrated into Egypt from Arabia. Among them were warriors and administrators, fine mechanics, artisans, artists and sculptors. They left us the Pyramids and other magnificent monumental tombs and great masses of architecture and sculptured columns. Of course, they declined and passed away, as all things human must; but they left behind them evidences to tell of their prestige and power.

The scientists and geologists of our day are busy unearthing the remains of the ancient peoples of the Eastern world, who started the waves of civilization both to the Orient and the Occident. Vast stores of knowledge are being accumulated and almost every day sees some ancient treasure trove brought to light. Especially in Biblical lands is the explorer busy unearthing the relics of the mighty past and throwing a flood of light upon incidents and scenes long covered by the dust of centuries.

Babylon, the mightiest city of ancient times, celebrated in the Bible and in the earliest human records as the greatest centre of sensual splendor and sinful luxury the world has ever seen, is at last being explored in the most thorough manner by the German Oriental Society, of which the Kaiser is patron. Babylon rose to its greatest glory under Nebuchadnezzar, the most famous monarch of the Babylonian Empire. At that period it was the great centre of arts, learning and science, astronomy and astrology being patronized by the Babylonian kings. The city finally came to a terrible end under Belshazzar, as related in the Bible. The palace of the impious king has been uncovered and its great piles of masonry laid bare. The great hall, where the young prophet Daniel read the handwriting on the wall, can now be seen. The palace stood on elevated ground and was of majestic dimensions. A winding chariot road led up to it. The lower part was of stone and the upper of burned bricks. All around on the outside ran artistic sculptures of men hunting animals. The doors were massive and of bronze and swung inward, between colossal figures of winged bulls. From the hall a stairway led to the throne room of the King, which was decorated with gold and precious stones and finished in many colors. The hall in which the infamous banquet was held was 140 feet by 40 feet. For a ceiling it was spanned by the cedars of Lebanon which exhaled a sweet perfume. At night a myriad lights lent brilliancy to the scene. There were over 200 rooms all gorgeously furnished, most of them devoted to the inmates of the king's harem. The ruins as seen to-day impress the visitor and excite wonder and admiration.

The Germans have also uncovered the great gate of Ishtar at Babylon, which Nebuchadnezzar erected in honor of the goddess of love and war, the most renowned of all the mythical deities of the Babylonian Pantheon. It is a double gateway with interior chambers, flanked by massive towers and was erected at the end of the Sacred Road at the northeast corner of the palace. Its most unique feature consists in the scheme of decoration on its walls, which are covered with row upon row of bulls and dragons represented in the brilliant enamelled bricks. Some of these creatures are flat and others raised in relief. Those in relief are being taken apart to be sent to Berlin, where they will be again put together for exhibition.

The friezes on this gate of Ishtar are among the finest examples of enamelled brickwork that have been uncovered and take their place beside "the Lion Frieze" from Sargon's palace at Khorsabad and the still more famous "Frieze of Arches of King Darius" in the Paris Louvre.

The German party have already established the claim of Herodotus as to the thickness of the walls of the city. Herodotus estimated them at two hundred royal cubits (348 feet) high and fifty royal cubits (86-1/2 feet) thick. At places they have been found even thicker. So wide were they that on the top a four-horse chariot could easily turn.

The hanging gardens of Babylon, said to have been built to please Amytis the consort of Nebuchadnezzar, were classed as among the Seven Wonders of the World. Terraces were constructed 450 feet square, of huge stones which cost millions in that stoneless country. These were supported by countless columns, the tallest of which were 160 feet high. On top of the stones were layers of brick, cemented and covered with pitch, over which was poured a layer of lead to make all absolutely water-tight. Finally, on the top of this, earth was spread to such a depth that the largest trees had room for their roots. The trees were planted in rows forming squares and between them were flower gardens. In fact, these gardens constituted an Eden in the air, which has never since been duplicated.

New discoveries have been recently made concerning the Tower of Babel, the construction of which, as described in the Book of Genesis, is one of the most remarkable occurrences of the first stage of the world's history. It has been found that the tower was square and not round, as represented by all Bible illustrators, including Dore. The ruins cover a space of about

50,000 square feet and are about ten miles from the site of Babylon.

The ruins of the celebrated synagogue of Capernaum, believed to be the very one in which the Saviour preached, have been unearthed and many other Biblical sites around the ancient city have been identified.

Capernaum was the home of Jesus during nearly the whole of his Galilean ministry and the scene of many of his most wonderful miracles. The site of Capernaum is now known as Tell Hum. There are ruins scattered about over a radius of a mile. The excavating which revealed the ruins of the synagogue was done under supervision of a German archaeologist named Kohl. This synagogue was composed of white limestone blocks brought from a distance and in this respect different from the others which were built of the local black volcanic rock. The carvings unearthed in the ruins are very beautiful and most of them in high relief work, representing trailing vines, stately palms, clusters of dates, roses and acanthus. Various animal designs are also shown and one of the famous seven-branched candlesticks which accompanied the Ark of the Covenant.

Most of the incidents at Capernaum mentioned in the Bible were connected with the synagogue, the ruins of which have just been uncovered. The centurion who came to plead with Jesus about the servant was the man who built the synagogue (Luke VII:1-10). In the synagogue, Jesus healed the man with the unclean spirit (Mark I:21-27). In this synagogue, the man with the withered hand received health on the Sabbath Day (Matthew XII:10-13). Jairus, whose daughter was raised from the dead, was a ruler of the synagogue (Luke VIII:3) and it was in this same synagogue of Capernaum that Jesus preached the discourse on the bread of life (John VI:26-59). The hill near Capernaum where Jesus fed the multitude with five loaves and two fishes is also identified.

The stoning of St. Stephen and the conversion of St. Paul are two great events of the New Testament which lend additional interest to the explorations now being carried on at the ancient City of Damascus. Damascus lays claim to being the most ancient city in the world and its appearance sustains the claim. Unlike Jerusalem and many other ancient cities, it has never been completely destroyed by a conqueror. The Assyrian monarch, Tiglath Pileser, swept down on it, 2,700 years ago, but he did not succeed in wiping it out. Other cities came into being long after Damascus, they flourished, faded and passed away; but Damascus still remains much the same as in the early time. Among the famous places which have been identified in this ancient city is the house of Ananias the priest and the place in the wall where Paul was let down by a basket is pointed out. The scene of the conversion of St. Paul is shown and also the "Street called Straight" referred to in Acts IX:II.

Jerusalem, birthplace and cradle of Christianity, offers a vast and interesting field to the archaeologist. One of the most remarkable of recent discoveries relates to the building known as David's castle. Major Conder, a British engineer in charge of the Palestine survey, has proved that this building is actually a part of the palace of King Herod who ordered the Massacre of the Innocents in order to encompass the destruction of the Infant Saviour.

The tomb of Hiram is another relic discovered at the village of Hunaneh on the road from Safed to Tyre; it recalls the days of David. Hiram was King of Tyre in the time of David. The tomb is a limestone structure of extraordinary massiveness Unfortunately the Mosque of Omar stands on the site of Solomon's Temple and there is no hope of digging there. As for the palace of Solomon, it should be easy to find the foundations, for Jerusalem has been rebuilt several times upon the ruins of earlier periods and vast ancient remains must be still buried there. The work is being pushed vigorously at present and the future should bring to light many interesting relics. At last the real site of the Crucifixion may be found with many mementoes of the Saviour, and the Apostles.

Professor Flinders Petrie, the famous English archaeologist, has recently explored the Sinaitic peninsula and has found many relics of the Hebrews' passage through the country during the Exodus and also many of a still earlier period. He found a remarkable number of altars and tombs belonging to a very early form of religion. On the Mount where Moses received the tables of the law is a monastery erected by the Emperor Justinian 523 A.D. Although the conquering wave of Islam has swept over the peninsula, leaving it bare and desolate, this monastery still survives, the only Christian landmark, not only in Sinai but in all Arabia. The original tables of stone on which the Commandments were written, were placed in the Ark of the Covenant and taken all through the Wilderness to Palestine and finally placed in the Temple of Solomon. What became of it when the Temple was plundered and destroyed by the Babylonians is not known.

Clay tablets have been found at Nineveh of the

Creation and the Flood as known to the Assyrians. These tablets formed part of a great epic poem of which Nimrod, the mighty hunter, was the hero.

Explorers are now looking for the palace of Nimrod, also that of Sennacherib, the Assyrian monarch who besieged Jerusalem. The latter despoiled the Temple of many of its treasures and it is believed that his palace, when found, may reveal the Tables of the Law, the Ark of the Covenant, the Seven-branched candlestick, and many of the golden vessels used in Israelitish worship.

Ur of the Chaldees, birthplace of Abraham, father and founder of the Hebrew race, is a rich field for the archaeologist to plough. Some tablets have already been discovered, but they are only a mere suggestion as to future possibilities. It is believed by some eminent investigators that we owe to Abraham the early part of the Book of Genesis describing the Creation, the Tower of Babel and the Flood, and the quest of archaeologists is to find, if not the original tablets, at least some valuable records which may be buried in this neighborhood.

Excavators connected with the American School at Jerusalem are busy at Samaria and they believe they have uncovered portions of the great temple of Baal, which King Ahab erected in honor of the wicked deity 890 B.C. When the remains of this temple are fully uncovered it will be learned just how far the Israelites forsook the worship of the true God for that of Baal.

The Germans have begun work on the site of Jericho, once the royal capital of Canaan, and historic chiefly from the fact that Joshua led the Israelites up to its walls, reported to be impregnable, but which "fell down at the blast of the trumpet." Great piles have been unearthed here which it is thought formed a part of the original masonry. One excavator believes he has unearthed the ruins of the house of Rahab, the woman who sheltered Joshua's spies. Another thinks he has discovered the site of the translation of Elijah, the Prophet, from whence he was carried up to heaven in a fiery chariot.

Every Christian will be interested in learning what is to be found in Nazareth where Jesus spent his boyhood. Archaeologists have located the "Fount of the Virgin," and the rock from which the infuriated inhabitants attempted to hurl Christ.

In the "Land of Goshen" where the Israelites in a state of servitude worked for the oppressing Pharaoh (Rameses II), excavators have found bricks made without straw as mentioned in Scripture, undoubtedly the work of Hebrew slaves, also glazed bead necklaces. They are looking for the House of Amran, the father of Moses, where the great leader was born.

The site of Arbela, where Alexander the Great won his mightiest victory over Darius, has been discovered. It is a series of mounds on the Western bank of the Tigris river between Nineveh and Bagdad. All the treasures of Darius were taken and Alexander erected a great palace. Bronze swords, cups and pieces of sculpture have been unearthed and it is supposed there are vast stores of other remains awaiting the tool and patience of the excavator. The famous Sultan Saladin took up his residence here in 1184 and doubtless many relics of his royal time will be discovered.

The remains of the city of Pumbaditha have been

identified with the immense mound of Abnar some twenty miles from Babylon, on the banks of the Euphrates. This was the centre of Jewish scholarship during the Babylonian exile. One of the great schools in which the Talmud was composed was located here. The great psalm, "By the waters of Babylon, we sat down and wept." was also composed on this spot, and here, too, Jeremiah and Isaiah thundered their impassioned eloquence. Broken tombs and a few inscribed bowls have been brought to light. Probably the original scrolls of the Talmud will be found here. Several curiously wrought vases and ruins have been also unearthed.

Several monuments bearing inscriptions which are sorely puzzling the archaeologists have recently been unearthed at the site of Boghaz-Keni which was the ancient, if not original capital, of the mysterious people called the Hittites who have been for so long a worry to Bible students. Archaeology has now revealed the secret of this people. There is no doubt they were of Mongolian origin, as the monuments just discovered represent them with slant eyes and pigtails. No one as yet has been able to read the inscriptions. They were great warriors, great builders and influenced the fate of many of the ancient nations.

In many other places throughout these lands, deep students of Biblical lore are pushing on the work of excavation and daily adding to our knowledge concerning the peoples and nations in whom posterity must ever take a vital interest.

A short time ago, Professor Doerpfeld announced to the world that he had discovered on the island of Ithaca, off the west coast of Greece, the ruins of the palace of Ulysses, Homer's half-mythical hero of the *Odyssey*. The German archaeologist has traced the different rooms of the palace and is convinced that here is the very place to which the hero returned after his wanderings. Near it several graves were found from which were exhumed silver amulets, curiously wrought necklaces, bronze swords and metal ornaments bearing date 2,000 B.C., which is the date at which investigators lay the Siege of Troy.

If the ruins be really those of the palace of Ulysses, some interesting things may be found to throw a light on the Homeric epic. As the schoolboys know, when Ulysses set sail from Troy for home, adverse winds wafted him to the coast of Africa and he beat around in the adjacent seas and visited islands and spent a considerable time meeting many kinds of curious and weird adventures, dallying at one time with the lotus-eaters, at another braving the Cyclops, the one-eyed monsters, until he arrived at Ithaca where "he bent his bow and slew the suitors of Penelope, his harassed wife."

In North America are mounds, earthworks, burial sites, shell heaps, buildings of stone and adobe, pictographs sculptured in rocks, stone implements, objects made of bone, pottery and other remains which arouse the enthusiasm of the archaeologist. As the dead were usually buried in America, investigators try to locate the ancient cemeteries because, besides skeletons, they usually contain implements, pottery and ornaments which were buried with the corpses. The most characteristic implement of early man in America was the grooved axe, which is not found in any other country. Stone implements are plentiful everywhere. Knives, arrow-points and perforators of chipped stone are found in all parts of the continent. Beads and shells and pottery are also found in almost every State.

The antiquity of man in Europe has been determined in a large measure by archaeological remains found in caves occupied by him in different ages, but the exploration of caves in North America has so far failed to reveal traces of different degrees of civilization.

CHAPTER IX

GREAT TUNNELS OF THE WORLD

Primitive Tunneling - Hoosac Tunnel - Croton Aqueduct - Great Alpine Tunnels - New York Subway - McAdoo Tunnels - How Tunnels are Built.

The art of tunnel construction ranks among the very oldest in the world, if not the oldest, for almost from the beginning of his advent on the earth man has been tunneling and boring and making holes in the ground. Even in pre-historic time, the ages of which we have neither record nor tradition, primitive man scooped out for himself hollows in the sides of hills, and mountains, as is evidenced by geological formations and by the fossils that have been unearthed. The forming of these hollows and holes was no indication of a superior intelligence but merely manifested the instincts of nature in seeking protection from the fury of the elements and safety from hostile forces such as the onslaughts of the wild and terrible beasts that then existed on the earth.

The Cave Dwellers were real tunnelers, inasmuch as in construction of their rude dwellings they divided them into several compartments and in most cases chose the base of hills for their operations, boring right through from side to side as recent discoveries have verified.

The ancient Egyptians built extensive tunnels for the tombs of their dead as well as for the temples of the living. When a king of Thebes ascended the throne he immediately gave orders for his tomb to be cut out of the solid rock. A separate passage or gallery led to the tomb along which he was to be borne in death to the final resting place. Some of the tunnels leading to the mausoleums of the ancient Egyptian kings were upwards of a thousand feet in length, hewn out of the hard solid rock. A similar custom prevailed in Assyria, Mesopotamia, Persia and India.

The early Assyrians built a tunnel under the Euphrates river which was 12 feet wide by 15 high. The course of the river was diverted until the tunnel was built, then the waters were turned into their former channel, therefore it was not really a subaqueous tunnel.

The sinking of tunnels under water was to be one of the triumphs of modern science.

Unquestionably the Romans were the greatest engineers of ancient times. Much of their masonry work has withstood the disintegrating hand of time and is as solid and strong to-day as when first erected.

The "Fire-setting" method of tunneling was originated by them, and they also developed the familiar principle of prosecuting the work at several points at the same time by means of vertical shafts. They heated the rock to be excavated by great fires built against the face of it. When a very high temperature was reached they turned streams of cold water on the heated stone with the result that great portions were disintegrated and fell off under the action of the water. The Romans being good chemists knew the effect of vinegar on lime, therefore when they encountered calcareous rock instead of water they used vinegar which very readily split up and disintegrated this kind of obstruction. The work of tunneling was very severe on the laborers, but the Romans did not care, for nearly all the workmen were slaves and regarded in no better light than so many cattle. One of the most notable tunnels constructed by the old Romans was that between Naples and Pozzuoli through the Posilipo Hills. It was excavated through volcanic tufa and was 3,000 feet long, 25 feet wide, and of the pointed arch style. The longest of the Roman tunnels, 3-1/2 miles, was built to drain Lake Fucino. It was driven through calcareous rock and is said to have cost the labor of 30,000 men for 11 years.

Only hand labor was employed by the ancient people in their tunnel work. In soft ground the tools used were picks, shovels and scoops, but for rock work they had a greater variety. The ancient Egyptians besides the hammer, chisel and wedges had tube drills and saws provided with cutting edges of corundum or other hard gritty material.

For centuries there was no progress in the art of tunneling. On the contrary there was a decline from the earlier construction until late in the 17th century when gunpowder came into use as an explosive in blasting rock. The first application of gunpowder was probably at Malpas, France, 1679-1681, in the construction of the tunnel on the line of the Languedoc Canal 510 feet long, 22 feet wide and 29 feet high.

It was not until the beginning of the nineteenth century

that the art of tunnel construction, through sand, wet ground or under rivers was undertaken so as to come rightly under the head of practical engineering. In 1803 a tunnel was built through very soft soil for the San Quentin Canal in France. Timbering or strutting was employed to support the walls and roof of the excavation as fast as the earth was removed and the masonry lining was built closely following it. From the experience gained in this tunnel were developed the various systems of soft ground subterranean tunneling in practice at the present day.

The first tunnel of any extent built in the United States was that known as the Auburn Tunnel near Auburn, Pa., for the water transportation of coal. It was several hundred feet long, 22 feet wide and 15 feet high. The first railroad tunnel in America was also in Pennsylvania on the Allegheny-Portage Railroad, built in 1818-1821. It was 901 feet long, 25 feet wide and 21 feet high.

What may be called the epoch making tunnel, the construction of which first introduced high explosives and power drills in this country, was the Hoosac in Massachusetts commenced in 1854 and after many interruptions brought to completion in 1876. It is a double-track tunnel nearly 5 miles in length. It was quickly followed by the commencement of the Erie tunnel through Bergen Hill near Hoboken, N.J. This tunnel was commenced in 1855 and finished in 1861. It is 4,400 feet long, 28 feet wide and 21 feet high. Other remarkable engineering feats of this kind in America are the Croton Aqueduct Tunnel, the Hudson River Tunnel, and the New York Subway.

The great rock tunnels of Europe are the four Alpine

cuts known as Mont Cenis, St. Gothard, the Arlberg and the Simplon. The Mont Cenis is probably the most famous because at the time of its construction it was regarded as the greatest engineering achievement of the modern world, yet it is only a simple tunnel 8 miles long, while the Simplon is a double tunnel, each bore of which is 12-1/4 miles. The chief engineer of the Mont Cenis tunnel was M. Sommeiler, the man who devised the first power drill ever used in such work. In addition to the power drill the building of this tunnel induced the invention of apparatus to suck up foul air, the air compressor, the turbine and several other contrivances and appliances in use at the present time.

Great strides in modern tunneling developed the "shield" and brought metal lining into service. The shield was invented and first used by Sir M. I. Brunel, a London engineer, in excavating the tunnel under the River Thames, begun in 1825 and finished in 1841. In 1869 another English engineer, Peter Barlow, used an iron lining in connection with a shield in driving the second tunnel under the Thames at London. From a use of the shield and metal lining has grown the present system of tunneling which is now universally known as the shield system.

Great advancement has been made in the past few years in the nature and composition of explosives as well as in the form of motive power employed in blasting. Powerful chemical compositions, such as nitroglycerine and its compounds, such as dynamite, etc., have supplanted gunpowder, and electricity, is now almost invariably the firing agent. It also serves many other purposes in the work, illumination, supplying power for hoisting and excavating machinery, driving rock drills, and operating ventilating fans, etc.

In this field, in fact, as everywhere else in the mechanical arts, the electric current is playing a leading part.

To the English engineer, Peter Barlow, above mentioned, must be given the credit of bringing into use the first really serviceable circular shield for soft ground tunneling. In 1863 he took out a patent for such a shield with a cylindrical cast iron lining for the completed tunnel. Of course James Henry Greathead very materially improved the shield, so much so indeed that the present system of tunneling by means of circular shields is called the Greathead not the Barlow system. Greathead and Barlow entered into a partnership in 1869. They constructed the tunnel under the Tower of London 1,350 feet in length and seven feet in diameter which penetrated compact clay and was completed within a period of eleven months. This was a remarkable record in tunnel building for the time and won for these eminent engineers a world wide fame. From thenceforth their system came into vogue in all soft soil and subaqueous tunneling. Except for the development in steel apparatus and the introduction of electricity as a motive agent, there has not been such a great improvement on the Greathead shield as one would naturally expect in thirty years.

The method of excavating a tunnel depends altogether on the nature of the obstruction to be removed for the passage. In the case of solid rock the work is slow but simple; dry, hard, firm earth is much the same as rock. The difficulties of tunneling lie in the soft ground, subaqueous mud, silt, quicksand, or any treacherous soil of a shifting, unsteady composition.

When the rock is to be removed it is customary to

begin the work in sections of which there may be seven or eight. First one section is excavated, then another and so on to completion. The order of the sections depends upon the kind of rock and upon the time allotted for the job and several other circumstances known to the engineer. If the first section attacked be at the top immediately beneath the arch of the proposed tunnel, next to the overlying matter, it is called a heading, but if the first cutting takes place at the bottom of the rock to form the base of the tunnel it is called a drift.

Driving a heading is the most difficult operation of rock tunneling. Sometimes a heading is driven a couple of thousand feet ahead of the other sections. In soft rock it is often necessary to use timber props as the work proceeds and follow up the excavating by lining roof and sides with brick, stone or concrete.

The rock is dislodged by blasting, the holes being drilled with compressed air, water force or electricity, and, as has been said, powerful explosives are used, nitroglycerine or some nitro-compound being the most common. Many charges can be electrically fired at the same time. If the tunnel is to be long, shafts are sunk at intervals in order to attack the work at several places at once. Sometimes these shafts are lined and left open when the tunnel is completed for purposes of ventilation.

In soft ground and subaqueous soil the "shield" is the chief apparatus used in tunneling. The most up-to-date appliance of this kind was that used in constructing the tunnels connecting New York City with New Jersey under the Hudson River. It consisted of a cylindrical shell of steel of the diameter of the excavation to be made. This was provided with a cutting edge of cast steel made up of assembled segments. Within the shell was arranged a vertical bulkhead provided with a number of doors to permit the passage of workmen, tools and explosives. The shell extended to the rear of the bulkhead forming what was known as the "tail." The lining was erected within this tail and consisted of steel plates lined with masonry. The whole arrangement was in effect a gigantic circular biscuit cutter which was forced through the earth.

The tail thus continually enveloped the last constructed portion of this permanent lining. The actual excavation took place in advance of the cutting edge. The method of accomplishing this, varied with conditions. At times the material would be rock for a few feet from the bottom, overlaid with soft earth. In such case the latter would be first excavated and then the roof would be supported by temporary timbers, after which the rock portion would be attacked. When the workmen had excavated the material in front of the shield it was passed through the heavy steel plate diaphragm in center of the shell out to the rear and the shield was then moved forward so as to bring its front again up to the face of the excavation. As the shell was very unwieldy, weighing about eighty tons, and, moreover, as the friction or pressure of the surrounding material on its side had to be overcome it was a very difficult matter to move it forward and a great force had to be expended to do so. This force was exerted by means of hydraulic jacks so devised and placed around the circumference of the diaphragm as to push against the completed steel plate lining of the tunnel. There were sixteen of these jacks employed with cylinders eight inches in diameter and they exerted a pressure of from one thousand to four thousand pounds per square inch.

By such means the shield was pushed ahead as soon as room was made in front for another move.

The purpose of the shield is to prevent the inrush of water and soft material while excavating is going on; the diaphragm of the shields acts as a bulkhead and the openings in it are so devised as to be quickly closed if necessary. The extension of the shield in front of the diaphragm is designed to prevent the falling or flowing in of the exposed face of the new excavation.

The extension of the shell back from the diaphragm is for the purpose of affording opportunity to put in place the finished tunnel lining whatever it may be, masonry, cast-iron, cast-iron and masonry, or steel plates and masonry. Where the material is saturated with water as is the case in all subaqueous tunneling it is necessary to use compressed air in connection with the shield. The intensity of air pressure is determined by the depth of the tunnel below the surface of the water above it. The tunnelers work in what are called caissons to which they have access through an air lock. In many cases quick transition from the compressed air in the caisson to the open air at the surface results fatally to the workers. The caisson disease is popularly called "the bends" a kind of paralysis which is more or less baffling to medical science. Some men are able to bear a greater pressure than others. It depends on the natural stamina of the worker and his state of health. The further down the greater the pressure. The normal atmospheric pressure at the surface is about fourteen pounds to the square inch. Men in normal health should be able to stand a pressure of seventy-six pounds to the square inch and this would call for a depth of 178 feet under water surface, which far exceeds any depth worked under compressed air. For a long time one hundred feet were regarded as a maximum depth and at that depth men were not permitted to work more than an hour in one shift. The ordinary subaqueous tunnel pressure is about forty-five pounds and this corresponds to a head of 104 feet. In working in the Hudson Tunnels the pressure was scarcely ever above thirty-three pounds, yet many suffered from the "bends."

What is called a freezing method is now proposed to overcome the water in soft earth tunneling. Its chief feature is the excavating first of a small central tunnel to be used as a refrigerating chamber or ice box in freezing the surrounding material solid so that it can be dug out or blasted out in chunks the same as rock. It is very doubtful however, if such a plan is feasible.

The greatest partly subaqueous tunnels in the world are now to be found in the vicinity of New York. The first to be opened to the public is known as the Subway and extends from the northern limits of the City in Westchester County to Brooklyn. The oldest, however, of the New York tunnels counting from its origin is the "McAdoo" tunnel from Christopher Street, in Manhattan Borough, under the Hudson to Hoboken. This was begun in 1880 and continued at intervals as funds could be obtained until 1890, when the work was abandoned after about two thousand feet had been constructed. For a number of years the tunnel remained full of water until it was finally acquired by the Hudson Companies who completed and opened it to the public in 1908. Another tunnel to the foot of Cortlandt Street was constructed by the same concern and opened in 1909. Both tunnels consist of parallel but separate tubes. The railway tunnels to carry the Pennsylvania R. R. under the Hudson into New York and thence under the East River to Long Island have been finished and are great triumphs of engineering skill besides making New York the most perfectly equipped city in the world as far as transit is concerned.

The greatest proposed subaqueous tunnel is that intended to connect England with France under the English Channel a distance of twenty-one miles. Time and again the British Parliament has rejected proposals through fear that such a tunnel would afford a ready means of invasion from a foreign enemy. However, it is almost sure to be built. Another projected British tunnel is one which will link Ireland and Scotland under the Irish Sea. If this is carried out then indeed the Emerald Isle will be one with Britain in spite of her unwillingness for such a close association.

England already possesses a famous subaqueous tunnel in that known as the Severn tunnel under the river of that name. It is four and a half miles long, although it was built largely through rock. Water gave much trouble in its construction which occupied thirteen years from 1873 to 1886. Pumps were employed to raise the water through a side heading connecting with a shaft twenty-nine feet in diameter. The greatest amount of water raised concurrently was twenty-seven million gallons in twenty-four hours but the pumps had a capacity of sixty-six million gallons for the same time.

The greatest tunnel in Europe is the Simplon which connects Switzerland with Italy under the Simplon Pass in the Alps. It has two bores twelve and one-fourth miles each and at places it is one and one-half miles below the surface. The St. Gothard also connecting Switzerland and Italy under the lofty peak of the Col de St. Gothard is nine and one-fourth miles in length. The third great Alpine tunnel, the Arlberg, which is six and one-half miles long, forms a part of the Austrian railway between Innsbruck and Bluedenz in the Tyrol and connects westward with the Swiss railroads and southward with those of Italy.

Two great tunnels at the present time are being constructed in the United States, one of these which is piercing the backbone of the Rockies is on the Atlantic and Pacific railway. It begins near Georgetown, will pass under Gray's peak and come out near Decatur, Colorado, in all a length of twelve miles. The other American undertaking is a tunnel under the famous Pike's Peak in Colorado which when completed will be twenty miles long.

It can clearly be seen that in the way of tunnel engineering Uncle Sam is not a whit behind his European competitors.

CHAPTER X

ELECTRICITY IN THE HOUSEHOLD

Electrically Equipped Houses - Cooking by Electricity - Comforts and Conveniences.

Science has now pressed the invisible wizard of electricity into doing almost every household duty from cleaning the windows to cooking the dinner. There are many houses now so thoroughly equipped with electricity from top to bottom that one servant is able to do what formerly required the service of several, and in some houses servants seem to be needed hardly at all, the mistresses doing their own cooking, ironing, and washing by means of electricity.

In respect to taking advantage of electricity to perform the duties of the household our friends in Europe were ahead of us, though America is pre-eminently the land of electricity - the natal home of the science. We are waking up, however, to the domestic utility of this agent and throughout the country at present there are numbers of homes in which electricity is employed to perform almost every task automatically from feeding the baby to the crimping of my lady's hair in her scented boudoir.

There is now no longer any use for chimneys on electrically equipped houses, for the fires have been eliminated and all heat and light drawn from the electric street mains. A description of one of these houses is most interesting as showing what really can be accomplished by this wonderful source of power.

Before the visitor to such a house reaches the gate or front door his approach is made known by an annunciator in the hall, which is connected with a hidden plate in the entrance path, which when pressed by the feet of the visitor charges the wire of the annunciator. A voice comes through the horn of a phonograph asking him what he wishes and telling him to reply through the telephone which hangs at the side of the door. When he has made his wants known, if he is welcome or desired, there is a click and the door opens. As he enters an electrically operated door mat cleans his shoes and if he is aware of the equipments of the house, he can have his clothes brushed by an automatic brush attached to the hat-rack in the hall. An escalator or endless stairway brings him to the first floor where he is met by the host who conducts him to the den sacred to himself. If he wishes a preprandial cigar, the host touches a segment of the wall, apparently no different in appearance from the surrounding surface, and a complete cigar outfit shoots out to within reach of the guest. When the gong announces dinner he is conducted to the dining hall where probably the uses to which electricity can be put are better exemplified than in any other part of the house. Between this room and the kitchen there is a perfect electric understanding. The apartments are so arranged that electric dumbwaiter service is operated between the centre of the dining table itself and the serving table in the kitchen. The latter is equipped with an electric range provided with electrically heated ovens, broilers, vegetable cookers, saucepans, dishes, etc., sufficient for the preparation of the most elaborate house banquet. The chef or cook in charge of the kitchen prepares each dish in its proper oven and has it ready waiting on the electric elevator at the appointed time when the host and his guest or guests, or family, as the case may be, are seated at the dining table. The host or whoever presides at the head of the table merely touches a button concealed on the side of the mahogany and the elevator instantly appears through a trap-door in the table, which is ordinarily closed by two silver covers which look like a tray. In this way the dish seemingly miraculously appears right on top of the table. When each guest is served it returns to the kitchen by the way it came and a second course is brought on the table in a similar manner and so on until the dinner is fully served. Fruits and flowers tastefully arranged adorn the centre of the dining table and minute electric incandescent lamps of various colors are concealed in the roses and petals and these give a very pretty effect, especially at night.

Beneath the table nothing is to be seen but two nickel-plated bars which serve to guide the elevators.

Down in the kitchen the cooking is carried on almost mechanically by means of an electric clock controlling the heating circuits to the various utensils. The cook, knowing just how long each dish will require to be cooked, turns on the current at the proper time and then sets the clock to automatically disconnect that utensil when sufficient time, so many minutes to the pound, has elapsed. When this occurs a little electric bell rings, calling attention to the fact, that the heat has been shut off.

Another kitchen accessory is a rotating table on which are mounted various household machines such as meat choppers, cream whippers, egg beaters and other apparatus all electrically operated.

There is also an electric dishwasher and dryer and plate rack manipulator which places the dishes in position when clean and dried.

The advantages of cooking by electricity are apparent to all who have tested them. Food cooked in an electric baking oven is much superior than when cooked by any other method because of the better heat regulation and the utter cleanliness, there being absolutely no dust whatever as in the case when coal is used. The electric oven does not increase the temperature nor does it exhaust the pure air in the room by burning up the oxygen. The time required for cooking is about the same as with coal.

The perfect cleanliness of an electric plate warmer is sufficient to warrant its use. It keeps dishes at a uniform temperature and the food does not get scorched and become tough.

Steaks prepared on electric gridirons and broilers are really delicious as they are evenly done throughout and retain all the natural juices of the meat; there is no odor of gas or of the fire and portions done to a crisp while others are raw on the inside. In toasting there is no danger of the bread burning on one side more than on the other, or of its burning on either side and a couple of dozen slices can be done together on an ordinary instrument at the same time. The electric diskstove, flat on the top, like a ball cut in two, can be also utilized as a toaster or for heating any kettles or pots or vessels with flat bottoms.

Very appetizing waffles are made with electric waffle irons, because the bottom and top irons are uniformly heated, so that the irons cook the waffles from both sides at the same time.

Electric potato peeling machines consist of a stationary cylinder opened at the top for the reception of the potatoes and having a revolving disk at the bottom. The cylinder has a rough surface or is coated with diamond flint, so that when the disk revolves the potatoes are thrown against the sides of the cylinder and the skin is scraped off. There is no deep cutting as when peeled by a knife, therefore, much waste is avoided. While the potatoes are being scraped, a stream of water plays upon them taking away the skins and thoroughly cleansing the tubers.

Among other electric labor savers connected with the culinary department may be mentioned floor-scrubbers, dish-washers, coffee-grinders, meat choppers, dough-mixers and cutlery-polishers, all of which give complete satisfaction at a paltry cost and save much time and labor. A small motor can drive any of these instruments or several can be attached and run by the same motor. The operation of an ordinary snap switch will supply energy to electric water-heaters attached to the kitchen boiler or to the faucet. The instantaneous water heater also purifies the water by killing the bacteria contained in it.

The electric tea kettle makes a brew to charm the heart of a connossieur. In fact all cooking done by electricity whether it is the frying of an egg or the roasting of a steak is superior in every way to the old methods and what accentuates its use is the cleanliness with which it can be performed. And it should be taken into consideration that in electric cooking there is no bending over hot stoves and ranges or a stuffy evil smelling smoky atmosphere, but on the contrary, fresh air, cleanliness and coolness which make cooking not the drudgery it has ever been, but a real pleasure.

Let us take a glance at the laundry in the electrically equipped house. There is a large tub with a wringer attached to it and a simple mechanism by which a small motor can either be connected with the tub or the wringer as required. The washing is performed entirely by the motor and in a way prevents the wear and tear associated with the old method of scrubbing and rubbing done at the expense of much "elbow grease." The motor turns the tub back and forth and in this way the soapy water penetrates the clothes, thus removing the dirt without injuring or tearing the fabric. In the old way, the clothes were moved up and down in the water and torn and worn in the process. By the new way it is the water which moves while the clothes remain stationary. When the clothes are thoroughly washed, the motor is attached to the wringer and they are passed through it; they are completely dried by a specially constructed electric fan. Whatever garments are to be ironed are separated and fed to a steel roll mangle operated by a motor which gives them a beautiful finish. The electric flat iron plays also an important part in the laundry as it is clean and never gets too hot nor too cold and there is no rushing back to replenish the heaters. One is not obliged to remain in the room with a hot stove, and suffer the inconveniences. No heat is felt at all from the iron as it is all concentrated on the bottom surface. It is a regular blessing to the laundress especially in hot weather.

There is a growing demand in all parts of the country for these electric flat-irons.

Electricity plays an important role in the parlor and drawing-room. The electric fireplace throws out a ruddy glow, a perfect imitation of the wide-open old-fashioned fireplaces of the days of our grandmothers. There are small grooves at certain sections in the flooring over which chairs and couches can be brought to a desired position. When the master drops into his favorite chair by the fireplace if he wishes a tune to soothe his jangled nerves, there is an electric attachment to the piano and he can adjust it to get the air of his choice without having to ask any one to play for him. In the drawing-room an electric fountain may be playing, its jets reflecting the prismatic colors of the rainbow as the waters fall in iridescent sparkle among the lights. Such a fountain is composed of a small electric motor and a centrifugal pump, the latter being placed in the interior of a basin and connected directly to the motor shaft. The pump receives the water from the basin and conveys it through pipes and a number of small nozzles thus producing cascades. The water falling upon an art glass dome, beneath which are small incandescent lamps, returns to the basin and thence again to the pump. There is no necessity of filling the fountain until the water gets low through evaporation. When the lights are not in colored glass, the water may be colored and this gives the same effect. To produce the play of the fountain and its effects, it is only necessary to connect it to any circuit and turn on the switch. The dome revolves by means of a jet of water driven against flanges on the under side of the rim of the dome and in this way beautiful and prismatic effects are produced. The motor is noiseless in operation. In addition to the pretty effect the fountain serves to cool and moisten the air of the room.

The sleeping chambers are thoroughly equipped. Not only the rooms may be heated by electricity but the beds themselves. An electric pad consisting of a flexible resistance covered with soft felt is connected by a conductor cord to a plug and is used for heating beds or if the occupant is suffering from rheumatism or indigestion or any intestinal pain this pad can be used in the place of the hot water bottle and gives greater satisfaction. There is a heat controlling device and the circuit can be turned on or off at will.

There are many more curious devices in the electrically equipped house which could they have been exhibited a generation or so ago, would have condemned the owner as a sorcerer and necromancer of the dark ages, but which now only place him in the category of the smart ones who are up to date and take advantage of the science and progress of the time.

CHAPTER XI

HARNESSING THE WATER-FALL

Electric Energy - High Pressure - Transformers - Development of Water-power.

The electrical transmission of power is exemplified in everything which is based on the generation of electricity. The ordinary electric light is power coming from a generator in the building or a public street-dynamo.

However, when we talk in general terms of electric transmission we mean the transmission of energy on a large scale by means of overhead or underground conductors to a considerable distance and the transformation of this energy into light and heat and chemical or mechanical power to carry on the processes of work and industry. When the power or energy is conveyed a long distance from the generator, say over 30 miles or more, we usually speak of the system of supply as long distance transmission of electric energy. In many cases power is conveyed over distances of 200 miles and more. When water power is available as at Niagara, the distance to which electric energy can be transmitted is considerably increased.

The distance to a great extent depends on the cost of coal required for generation at the distributing point and on the amount of energy demanded at the receiving point. Of course the farther the distance the higher must be the voltage pressure.

Electrical engineers say that under proper conditions electric energy may be transmitted in large quantity to a distance of 500 miles and more at a pressure of about 170,000 volts. If such right conditions be established then New York, Chicago and several other of our large cities can get their power from Niagara.

In our cities and towns where the current has only to go a short distance from the power house, the conductors are generally placed in cables underground and the maximum electro-motive force scarcely ever exceeds 11,000 volts. This pressure is generated by a steam-driven alternating-current generator and is transmitted over the conductors to sub-stations, where by means of step-down transformers, the pressure is dropped to, say, 600 volts alternating current which by rotary converters is turned into direct current for the street mains, for feeders of railways and for charging storage batteries which in turn give out direct current at times of heavy demand.

That electric transmission of energy to long distances may be successfully carried out transformers are necessary for raising the pressure on the transmission line and for reducing it at the points of distribution. The transformer consists of a magnetic circuit of laminated iron or mild steel interlinked with two electric circuits, one, the primary, receiving electrical energy and the other the secondary, delivering it to the consumer. The effect of the iron is to make as many as possible of the lines of force set up by the primary current, cut the secondary winding and there set up an electromotive force of the same frequency but different voltage.

The transformer has made long distance the actual achievement that it is. It is this apparatus that brought the mountain to Mohammed. Without it high pressure would be impossible and it is on high pressure that success of long distance transmission depends.

To convey electricity to distant centres at a low pressure would require thousands of dollars in copper cables alone as conductors. To illustrate the service of the transformer in electricity it is only necessary to consider water power at a low pressure. In such a case the water can only be transmitted at slow speed and through great openings, like dams or large canals, and withal the force is weak and of little practical efficiency, whereas under high pressure a small quantity can be forced through a small pipe and create an energy beyond comparison to that developed when under low pressure.

The transformer raises the voltage and sends the electrical current under high pressure over a small wire and so great is this pressure that thousands of horsepower can be sent to great distances over small wires with very little loss.

Water power is now changed to electrical power and transmitted over slender copper wires to the great manufacturing centres of our country to turn the wheels of industry and give employment to thousands.

Nearly one hundred cities in the United States alone

are today using electricity supplied by transmitted water-power. Ten years ago Niagara Falls were regarded only as a great natural curiosity of interest only to the sightseer, today those Falls distribute over 100,000 horse-power to Buffalo, Syracuse, Rochester, Toronto and several smaller cities and towns. Wild Niagara has at last indeed been harnessed to the servitude of man. Spier Falls north of Saratoga, practically unheard of before, is now supplying electricity to the industrial communities of Schenectady, Troy, Amsterdam, Albany and half a dozen or so smaller towns.

Rivers and dams, lakes and falls in all parts of the country are being utilized to supply energy, though at the present time only about one-fortieth of the horse-power available through this agent is being made productive. The water conditions of the United States are so favorable that 200,000,000 horse-power could be easily developed, but as it is we have barely enough harnessed to supply 5 million horse-power.

Eighty per cent. of the power used at the present time is produced from fuel. This percentage is sure to decrease in the future for fuel will become scarcer and the high cost will drive fuel power altogether out of the market.

New York State has the largest water power development in the Union, the total being 885,862 horsepower; this fact is chiefly owing to the energy developed by Niagara.

The second State in water-power development is California, the total development being 466,774 horsepower over 1,070 wheels or a unit installation of about 436 H.P.

The third State is Maine with 343,096 horse-power, over 2,707 wheels or an average of about 123 horse-power per wheel.

Lack of space makes it impossible to enter upon a detailed description of the structural and mechanical features of the various plants and how they were operated for the purpose of turning water into an electric current. The best that can be done is to outline the most noteworthy features which typify the various situations under which power plants are developed and operated.

The water power available under any condition depends principally upon two factors: First, the amount of fall or hydrostatic head on the wheels; second, the amount of water that can be turned over the wheels. The conditions vary according to place, there are all kinds of fall and flow. To develop a high power it is necessary to discharge a large volume of water upon properly designed wheels. In many of the western plants where only a small amount of water is available there is a great fall to make up for the larger volume in force coming down upon the wheels. So far as actual energy is concerned it makes no difference whether we develop a certain amount of power by allowing twenty cubic feet of water per second to fall a distance of one foot or allow one cubic foot of water per second to fall a distance of twenty feet.

In one place we may have a plant developing say 10,000 horse-power with a fall of anywhere from twenty to forty feet and in another place a plant of the same capacity with a fall of 1,000, 1,500 or 2,000 feet.

In the former case the short fall is compensated by a great volume of water to produce such a horse-power, while in the latter converse conditions prevail. In many cases the power house is located some distance from the source of supply and from the point where the water is diverted from its course by artificial means.

The Shawinigan Falls of St. Maurice river in Canada occur at two points a short distance apart, the fall at one point being about 50 and at the other 100 feet high. A canal 1,000 feet long takes water from the river above the upper of these falls and delivers it near to the electric power house on the river bank below the lower falls. In this way a hydrostatic head of 125 feet is obtained at the power house. The canal in this case ends on high ground 130 feet from the power house and the water passes down to the wheels through steel penstocks 9 feet in diameter.

In a great many cases in level country the water power can only be developed by means of such canals or pipe lines and the generating stations must be situated away from the points where the water is diverted from its course.

In mountainous country where rivers are comparatively small and their courses are marked by numerous falls and rapids, it is generally necessary to utilize the fall of a stream through some miles of its length in order to get a satisfactory development of power. To reach this result rather long canals, flumes, or pipe lines must be laid to convey the water to the power stations and deliver it at high pressure.

California offers numerous examples of electric power development with the water that has been carried several miles through artificial channels. An illustration of this class of work exists at the electric power house on the bank of the Mokelumne river in the Sierra Nevada mountains. Water is supplied to the wheels in this station under a head of 1,450 feet through pipes 3,600 feet long leading to the top of a near-by hill. To reach this hill the water after its diversion from the Mokelumne river at the dam, flows twenty miles through a canal or ditch and then through 3,000 feet of wooden stave pipe. Although California ranks second in water-power development it is easily the first in the number of its stations, and also be it said, California was the first to realize the possibilities of long distance electrical energy. The line from the 15,000 horsepower plant at Colgate in this State to San Francisco by way of Mission San Jose, where it is supplied with additional power, has a length of 232 miles and is the longest transmission of electrical energy in the world. The power house at Colgate has a capacity of 11,250 kilowatts in generators, but it is uncertain what part of the output is transmitted to San Francisco, as there are more than 100 substations on the 1,375 miles of circuit in this system.

Another system, even greater than the foregoing which has just been completed is that of the Stanislaus plant in Tuolumme County, California, from which a transmission line on steel towers has been run in Tuolumme, Calaveras, San Joaquin, Alameda and Contra Costa Counties for the delivery of power to mines and to the towns lying about San Francisco Bay. The rushing riotous waters of the Stanislaus wasted for so many centuries have been saved by the steel paddles of gigantic turbine water wheels and converted into electricity which carries with the swiftness of thought thousands of horse power energy to the far away cities and towns to be transformed into light and heat and power to run street cars and trains and set in motion the mechanism of mills and factories and make the looms of industry hum with the bustle and activity of life.

It is said that the greatest long distance transmission yet attempted will shortly be undertaken in South Africa where it is proposed to draw power from the famous Victoria Falls. The line from the Falls will run to Johannesburg and through the Rand, a length of 700 miles. It is claimed the Falls are capable of developing 300,000 electric horse power at all times.

Should this undertaking be accomplished it will be a crowning achievement in electrical science.

CHAPTER XII

WONDERFUL WARSHIPS

Dimensions, Displacements, Cost and Description of Battleships - Capacity and Speed - Preparing for the Future.

All modern battleships are of steel construction. The basis of all protection on these vessels is the protective deck, which is also common to the armored cruiser and many varieties of gunboats. This deck is of heavy steel covering the whole of the vessel a little above the water-line in the centre; it slopes down from the centre until it meets the sides of the vessel about three feet below the water; it extends the entire length of the ship and is firmly secured at the ends to the heavy stem and stern posts. Underneath this deck are the essentials of the vessel, the boilers and machinery, the magazines and shell rooms, the ammunition cells and all the explosive paraphernalia which must be vigilantly safeguarded against the attacks of the enemy. Every precaution is taken to insure safety. All openings in the protective deck above are covered with heavy steel gratings to prevent fragments of shell or other combustible substances from getting through to the magazine or powder cells.

The heaviest armor is usually placed at the water line because it is this part of the ship which is the most vulnerable and open to attack and where a shell or projectile would do the most harm. If a hole were torn in the side at this place the vessel would quickly take in water and sink. On this account the armor is made thick and is known as the water-line belt. At the point where the protective deck and the ship's side meet, there is a projection or ledge on which this armor belt rests. Thus it goes down about three feet below the water and it extends to the same distance above.

The barbettes, that is, the parapets supporting the gun turrets, are one forward and one aft. They rest upon the protective deck at the bottom and extend up about four feet above the upper deck. At the top of the barbettes, revolving on rollers, are the turrets, sometimes called the hoods, containing the guns and the leading mechanism and all of the machinery in connection with the same. The turret ammunition hoists lead up from the magazine below, delivering the charges and projectiles for the guns at the very breach so that they can be loaded immediately.

An athwartship line of armor runs from the water line to the barbettes, resting upon the protective deck. In fact, the space between the protective and upper deck is so closed in with armor, with a barbette at each end, that it is like a citadel or fort or some redoubt well-guarded from the enemy. Resting upon the water-belt and the athwartship or diagonal armor, and following the same direction is a layer of armor usually somewhat thinner which is called the lower case-mate armor; it extends up to the lower edge of the broadside gun ports, and resting upon it in turn is the upper case-mate armor, following the same direction, and forming the protection for the broadside battery. The explosive effect of the modern shell is so tremendous that were one to get through the upper case-mate and explode immediately after entering, it would undoubtedly disable several guns and kill their entire crews; it is, therefore, usual to isolate each broadside gun from its neighbors by light nickel steel bulkheads a couple of inches or so thick, and to prevent the same disastrous result among the guns on the opposite side, a fore-and-aft bulkhead of about the same thickness is placed on the centre line of the ship. Each gun of the broadside battery is thus mounted in a space by itself somewhat similar to a stall. Abaft the forward turret there is a vertical armored tube resting on the protective deck and at its upper end is the conning tower, from which the ship is worked when in action and which is well safe-guarded.

The tube protects all the mechanical signalling gear running into the conning tower from which communication can be had instantly with any part of the vessel.

To build a battleship that will be practically unsinkable by the gun fire of an enemy it is only necessary to make the water belt armor thick enough to resist the shells, missiles and projectiles aimed at it. There is another essential that is equally important, and that is the protection of the batteries. The experience of modern battles has made it manifest, that it is impossible for the crew to do their work when exposed to a hail of shot and shell from a modern battery of rapid fire and automatic guns. And so in all more recently built battleships and armored cruisers and gunboats, the protection of broadside batteries and exposed positions has been increased even at the expense of the water-line belt.

Armor plate has been much improved in recent years. During the Civil War the armor on our monitors was only an inch thick. Through such an armor the projectiles of our time would penetrate as easily as a bullet through a pine board. It was the development of gun power and projectiles that called forth the thick armor, but it was soon found that it was impossible for the armor to keep pace with the deadliness of the guns as it was utterly impossible to carry the weight necessary to resist the force of impact. Then came the use of special plates, the compound armor where a hard face to break up the projectile was welded to a softer back to give the necessary strength. This was followed by the steel armor treated by the Harvey process; it was like the compound armor in having a hard face and a soft back, but the plates were made from a single ingot without any welding.

The Harvey process enabled an enormously greater resistance to be obtained with a given weight of armor, but even it has been surpassed by the Krupp process which enables twelve inches of thickness to give the same resistance as fifteen of Harveyized plates.

The armament or battery of warships is divided into two classes, viz., the main and the second batteries. The main battery comprises the heaviest guns on the ship, those firing large shell and armor-piercing projectiles, while the second battery consists of small rapid fire and machine guns for use against torpedo boats or to attack the unprotected or lightly protected gun positions of an enemy. The main battery of our modern battleships consists usually of ten twelve-inch guns, mounted in pairs on turrets in the centre of the ship. In addition to these heavy guns it is usual to mount a number of smaller ones of from five to eight inches diameter of bore on each broadside, although sometimes they are mounted on turrets like the larger guns.

A twelve-inch breech-loading gun, fifty calibers long and weighing eighty-three tons, will propel a shell weighing eight hundred and eighty pounds, by a powder charge of six hundred and twenty-four pounds, at a velocity of over two thousand six hundred and twenty feet per second, giving an energy at the muzzle of over forty thousand foot-tons and is capable of penetrating at the muzzle, forty-five inches of iron.

During the last few years, very large increases have been made in the dimensions, displacements and costs of battleships and armored cruisers as compared with vessels of similar classes previously constructed. Both England and the United States have constructed enormous war vessels within the past decade. The British *Dreadnought* built in nineteen hundred and five has a draft of thirty-one feet six inches and a displacement of twenty-two thousand and two hundred tons. Later, vessels of the *Dreadnought* type have a normal draft of twenty-seven feet and a naval displacement of eighteen thousand and six hundred tons. Armored cruisers of the British *Invincible* class have a draft of twenty-six feet and a displacement of seventeen thousand two hundred and fifty tons with a thousand tons of coal on board. These cruisers have engines developing forty-one thousand horse-power.

Within the past two years the United States has turned out a few formidable battleships, which it is claimed surpass the best of those of any other navy in the world. The *Delaware* and *North Dakota* each have a draft of twenty-six feet, eleven inches and a displacement of twenty thousand tons. Great interest attached to the trials of these vessels because they were sister ships fitted with different machinery and it was a matter of much speculation which would develop the greater speed. In addition to the consideration of the battleship as a fighting machine at close quarters, Uncle Sam is trying to have her as fleet as an ocean greyhound should an enemy heave in sight so that the latter would not have much opportunity to show his heels to a broadside. The *Delaware*, which has reciprocating engines, exceeded her contract speed of twenty-one knots on her runs over a measured mile course in Penobscot Bay on October 22 and 23, 1909. Three runs were made at the rate of nineteen knots, three at 20.50 knots, and five at 21.98 knots.

The *North Dakota* is furnished with Curtis turbine engines. Here is a comparison of the two ships:

	North Dakota	Delaware
Fastest run over measured mile	21.98	22.25
Average of five high runs	21.44	21.83
Full power trial speed	21.56	21.64
Full power trial horsepower	28,600	31,400.
Full power trial, coal consumption, tons per day	578.	583.
Nineteen-knot trial coal consumption, tons per day	315.	295.
Twelve-knot trial coal consumption, tons per day	111.	105.

The *Florida*, a 21,825 ton boat, was launched from the Brooklyn Navy Yard last May 12. Her sister ship, the *Utah*, took water the previous December at Camden.

Here is a comparison of the *North Dakota* of 1908 and the *Florida* of 1910:

	N. Dakota	Florida
Length	518 ft. 9 in.	521 ft. 6 in.
Beam	85 ft. 2-1/2 in.	88 ft. 2-1/2 in.
Draft, Mean	26 ft. tons	21,825 tons
Coal Supply	2,500 tons	2,500 tons
Oil	400 tons	400 tons
Belt Armor	12 in. to 8 in.	12 in. to 8 in.
Turret Armor	12 11 in.	28 ft. 6 in.
Displacement	20,000 inches	12 inches
Battery armor	6 in.	6-1/2 in.
Smoke stack protection	6 inches	9-1/2 inches
12-inch guns	Ten	Ten
5-inch guns	Fourteen	Sixteen
Speed	21 knots	20.75 knots

The *Florida* has Parsons turbines working on four shafts and generates 28,000 horse-power.

The United States Navy has planned to lay down next year (1911) two ships of 32,000 tons armed with 14-inch guns, each to cost eighteen million dollars as compared with the $11,000,000 ships of 1910.

The following are to be some of the features of the projected ships, which are to be named the *Arkansas* and *Wyoming*.

554 ft. long, 93 ft. 3 in. beam, 28 ft. 6 in. draft, 26,000 tons displacement, 28,000 horse-power, 30 1/2 knots speed, 1,650 to 2,500 tons coal supply, armament of twelve 12-inch guns, twenty-one 5-inch, four 3-pounders and two torpedo tubes.

Fittings in recent United States battleships are for

21-inch torpedoes. The armor is to be 11 inch on belt and barbettes and on sides 8 inches, and each ship is to carry a complement of 1,115 officers and men. Two of the turrets will be set forward on the forecastle deck, which will have 28 feet, freeboard, the guns in the first turret being 34 feet above the water and those of the second about 40 feet. Aft of the second turret will be the conning tower, and then will come the fore fire-control tower or lattice mast, with searchlight towers carried on it. Next will come the forward funnel, on each side of which will be two small open rod towers with strong searchlights. Then will come the main fire-control tower and the after funnel and another open tower with searchlight. The two lattice steel towers are to be 120 feet high and 40 feet apart. The four remaining turrets will be abaft the main funnel, the third turret having its guns 32 feet above water; those in the other turrets about 25 feet above the water. The guns will be the new 50-calibre type. All twelve will have broadside fire over a wide arc and four can be fired right ahead and four right astern.

CHAPTER XIII

A TALK ON BIG GUNS

The First Projectiles - Introduction of Cannon - High Pressure Guns - Machine Guns - Dimensions and Cost of Big Guns.

The first arms and machines employing gunpowder as the propelling agency, came into use in the fourteenth century. Prior to this time there were machines and instruments which threw stones and catapults and large arrows by means of the reaction of a tightly twisted rope made up of hemp, catgut or hair. Slings were also much employed for hurling missiles.

The first cannons were used by the English against the Scots in 1327. They were short and thick and wide in the bore and resembled bowls or mortars; in fact this name is still applied to this kind of ordnance. By the end of the fifteenth century a great advancement was shown in the make of these implements of warfare. Bronze and brass as materials came into general use and cannon were turned out with twenty to twenty-five inch bore weighing twenty tons and capable of hurling to a considerable distance projectiles weighing from two hundred pounds to one thousand pounds with powder as the propelling force. In a short time these large guns were mounted and carriages were introduced to facilitate transportation with troops. Meantime stone projectiles were replaced by cast iron shot, which, owing to its greater density, necessitated a reduction in calibre, that is a narrowing of the bore, consequently lighter and smaller guns came into the field, but with a greater propelling force. When the cast iron balls first came into use as projectiles, they weighed about twelve pounds, hence the cannons shooting them were known as twelve-pounders. It was soon found, however, that twelve pounds was too great a weight for long distances, so a reduction took place until the missiles were cut down to four pounds and the cannon discharging these, four pounders as they were called, weighed about one-quarter of a ton. They were very effective and handy for light field work.

The eighteenth century witnessed rapid progress in gun and ammunition manufacture. "Grape" and "canister" were introduced and the names still cling to the present day. Grape consisted of a number of tarred lead balls, held together in a net. Canister consisted of a number of small shot in a tin can, the shots being dispersed by the breaking of the can on discharge. Grape now consists of cast iron balls arranged in three tiers by means of circular plates, the whole secured by a pin which passes through the centre. The number of shot in each tier varies from three to five. Grape is very destructive up to three hundred yards and effective up to six hundred yards. Canister shot as we know it at present, is made up of a number of iron balls, placed in a tin cylinder with a wooden bottom, the size of the piece of ordnance for which it is intended.

Towards the close of the eighteenth century, short cast-iron guns called "carronades" were introduced by

Gascoigne of the Cannon Iron Works, Scotland. They threw heavy shots at low velocity with great battery effect. They were for a long time in use in the British navy. The sailors called them "smashers."

The entire battery of the Victory, Nelson's famous flag-ship at the battle of Trafalgar, amounting to a total of 102 guns, was composed of "carronades" varying in size from thirty-two to sixty-eight pounders. They were mounted on wooden truck carriages and were given elevation by handspikes applied under the breech, a quoin or a wedge shaped piece of wood being pushed in to hold the breech up in position. They were trained by handspikes with the aid of side-tackle and their recoil was limited by a stout rope, called the breeching, the ends of which were secured to the sides of the ship. The slow match was used for firing, the flint lock not being applied to naval guns until 1780.

About the middle of the nineteenth century, the design of guns began to receive much scientific thought and consideration. The question of high velocities and flat trajectories without lightening the weight of the projectile was the desideratum; the minimum of weight in the cannon itself with the maximum in the projectile and the force with which it could be propelled were the ends to be attained.

In 1856 Admiral Dahlgren of the United States Navy designed the *Dahlgren* gun with shape proportioned to the "curve of pressure," which is to say that the gun was heavy at the breech and light at the muzzle. This gun was well adapted to naval use at the time. From this, onward, guns of high pressure were manufactured until the pressure grew to such proportions that it exceeded the resisting power, represented by the tensile strength of cast iron. When cast, the gun cooled from the outside inwardly, thus placing the inside metal in a state of tension and the outside in a state of compression. General Rodman, Chief of Ordnance of the United States Army, came forward with a remedy for this. He suggested the casting of guns hollow and the cooling of them from the inside outwardly by circulating a stream of cold water in the bore while the outside surface was kept at a high temperature. This method placed the metal inside in a state of compression and that on the outside in a state of tension, the right condition to withstand successfully the pressure of the powder gas, which tended to expand the inner portions beyond the normal diameter and throw the strain of the supporting outer layers.

This system was universally employed and gave the best results obtainable from cast iron for many years and was only superseded by that of "built up" guns, when iron and steel were made available by improved processes of production.

The great strides made in the manufacture and forging of steel during the past quarter of a century, the improved tempering and annealing processes have resulted in the turning out of big guns solely composed of steel.

The various forms of modern ordnance are classified and named according to size and weight, kind of projectiles used and their velocities; angle of elevation at which they are fired; use; and mode of operation.

The guns known as breechloading rifles are from three inches to fourteen inches in calibre, that is, across the bore, and in length from twelve to over sixty feet. They weigh from one ton to fifty tons.

They fire solid shot or shells weighing up to eleven hundred pounds at high velocities, from twenty-three to twenty-five hundred feet per second. They can penetrate steel armor to a depth of fifteen to twenty inches at 2,000 yards distance.

Rapid fire guns are those in which the operation of opening and closing the breech is performed by a single motion of a lever actuated by the hand, and in which the explosive used is closed in a metallic case. These guns are made in various forms and are operated by several different systems of breech mechanism generally named after their respective inventors. The Vickers-Maxim and the Nordenfeldt are the best known in America. A new type of the Vickers-Maxim was introduced in 1897 in which a quick working breech mechanism automatically ejects the primer and draws up the loading tray into position as the breech is opened. This type was quickly adopted by the United States Navy and materially increased the speed of fire in all calibres.

What are known as machine guns are rapid fire guns in which the speed of firing is such that it is practically continuous. The best known make is the famous Gatling gun invented by Dr. R. J. Gatling of Indianapolis in 1860. This gun consists of ten parallel barrels grouped around and secured firmly to a main central shaft to which is also attached the grooved cartridge carrier and the lock cylinder. Each barrel is provided with its own lock or firing mechanism, independent of the other, but all of them revolve simultaneously with the barrels, carrier and inner breech when the gun is in operation. In firing, one end of the feed case containing the cartridges is placed in the hopper on top and the operating crank is turned. The cartridges drop one by one into the grooves of the carrier and are loaded and fired by the forward motion of the locks, which also closes the breech while the backward motion extracts and expels the empty shells. In its present state of efficiency the Gatling gun fires at the rate of 1,200 shots per minute, a speed, by separate discharges, not equaled by any other gun.

Much larger guns were constructed in times past than are being built now. In 1880 the English made guns weighing from 100 to 120 tons, from 18 to 20 inches bore and which fired projectiles weighing over 2,000 pounds at a velocity of almost 1,700 feet per second. At the same time the United States fashioned a monster rifle of 127 tons which had a bore of sixteen inches and fired a projectile of 2,400 pounds with a velocity of 2,300 feet per second.

The largest guns ever placed on board ship were the Armstrong one-hundred-and-ten-ton guns of the English battleships, *Sanspareil*, *Benbow* and *Victoria*. They were sixteen and one-fourth inch calibre. The newest battleships of England, the *Dreadnought* and the *Temeraire*, are equipped with fourteen-inch guns, but they are not one-half so heavy as the old guns. Many experts in naval ordnance think it a mistake to have guns over twelve inch bore, basing their belief on the experience of the past which showed that guns of a less calibre carrying smaller shells did more effective work than the big bore guns with larger projectiles.

The two titanic war-vessels now in course of construction for the United States Navy will each carry a battery of ten fourteen-inch rifles, which will be the most powerful weapons ever constructed and will greatly exceed in range and hitting power the twelve-inch guns of the *Delaware* or *North Dakota*. Each of the new rifles will weigh over sixty-three tons, the projectiles will each weigh 1,400 pounds and the powder charge will be 450 pounds. At the moment of discharge each of these guns will exert a muzzle energy of 65,600 foot tons, which simply means that the energy will be so great that it could raise 65,600 tons a foot from the ground. The fourteen-hundred-pound projectiles shall be propelled through the air at the rate of half a mile a second. It will be plainly seen that the metal of the guns must be of enormous resistance to withstand such a force. The designers have taken this into full consideration and will see to it that the powder chamber in which the explosion takes place as well as the breech lock on which the shock is exerted is of steel so wrought and tempered as to withstand the terrific strain. At the moment of detonation the shock will be about equal to that of a heavy engine and a train of Pullman coaches running at seventy miles an hour, smashing into a stone wall. On leaving the muzzle of the gun the shell will have an energy equivalent to that of a train of cars weighing 580 tons and running at sixty miles an hour. Such energy will be sufficient to send the projectile through twenty-two and a half inches of the hardest of steel armour at the muzzle, while at a range of 3,000 yards, the projectile moving at the rate of 2,235 feet per second will pierce eighteen and a half inches of steel armor at normal impact. The velocity of the projectile leaving the gun will be 2,600 feet per second, a speed which if maintained would carry it around the world in less than fifteen hours.

Each of the mammoth guns will be a trifle over

fifty-three feet in length and the estimated cost of each will be $85,000. Judging from the performance of the twelve-inch guns it is figured that these greater weapons should be able to deliver three shots a minute. If all ten guns of either of the projected *Dreadnoughts* should be brought into action at one time and maintain the three shot rapidity for one hour, the cost of the ammunition expended in that hour would reach the enormous sum of $2,520,000.

Very few, however, of the big guns are called upon for the three shots a minute rate, for the metal would not stand the heating strain.

The big guns are expensive and even when only moderately used their "life" is short, therefore, care is taken not to put them to too great a strain. With the smaller guns it is different. Some of six-inch bore fire as high as eight aimed shots a minute, but this is only under ideal conditions.

Great care is being taken now to prolong the "life" of the big guns by using non-corrosive material for the charges. The United States has adopted a pure gun-cotton smokeless powder in which the temperature of combustion is not only lower than that of nitro-glycerine, but even lower than that of ordinary gunpowder. With the use of this there has been a very material decrease in the corrosion of the big guns. The former smokeless powder, containing a large percentage of nitro-glycerine such as "cordite," produced such an effect that the guns were used up and practically worthless, after firing fifty to sixty rounds.

Now it is possible for a gun to be as good after two or even three hundred rounds as at the beginning, but certainly not if a three minute rate is maintained. At such a rate the "life" of the best gun made would be short indeed.

CHAPTER XIV

MYSTERY OF THE STARS

Wonders of the Universe - Star Photography - The Infinity of Space.

In another chapter we have lightly touched upon the greatness of the Universe, in the cosmos of which our earth is but an infinitesimal speck. Even our sun, round which a system of worlds revolve and which appears so mighty and majestic to us, is but an atom, a very small one, in the infinitude of matter and as a cog, would not be missed in the ratchet wheel which fits into the grand machinery of Nature.

If our entire solar system were wiped out of being, there would be left no noticeable void among the countless systems of worlds and suns and stars; in the immensity of space the sun with all his revolving planets is not even as a drop to the ocean or a grain of sand to the composition of the earth. There are millions of other suns of larger dimensions with larger attendants wheeling around them in the illimitable fields of space. Those stars which we erroneously call "fixed" stars are the centers of other systems vastly greater, vastly grander than the one of which our earth forms so insignificant a part. Of course to us numbers of them appear, even when viewed through the most powerful telescopes, only as mere luminous points, but that is owing to the immensity of distance between them and ourselves. But the number that is visible to us even with instrumental assistance can have no comparison with the number that we cannot see; there is no limit to that number; away in what to us may be called the background of space are millions, billions, uncountable myriads of invisible suns regulating and illuminating countless systems of invisible worlds. And beyond those invisible suns and worlds is a region which thought cannot measure and numbers cannot span. The finite mind of man becomes dazed, dumbfounded in contemplation of magnitude so great and distance so amazing. We stand not bewildered but lost before the problem of interstellar space. Its length, breadth, height and circumference are illimitable, boundless; the great eternal cosmos without beginning and without end.

In order to get some idea of the vastness of interstellar space we may consider a few distances within the limits of human conception. We know that light travels at the rate of 186,000 miles a second, yet it requires light over four years to reach us from the nearest of the fixed stars, travelling at this almost inconceivable rate, and so far away are some that their light travelling at the same rate from the dawn of creation has never reached us yet or never will until our little globule of matter disintegrates and its particles, its molecules and corpuscles, float away in the boundless ether to amalgamate with the matter of other flying worlds and suns and stars.

The nearest to us of all the stars is that known as *Alpha Centauri*. Its distance is computed at

25,000,000,000,000 miles, which in our notation reads twenty-five trillion miles. It takes light over four years to traverse this distance. It would take the "Empire State Express," never stopping night or day and going at the rate of a mile a minute, almost 50,000,000 years to travel from the earth to this star. The next of the fixed stars and the brightest in all the heavens is that which we call *Sirius* or the Dog Star. It is double the distance of Alpha Centauri, that is, it is eight "light years" away. The distances of about seventy other stars have been ascertained ranging up to seventy or eighty "light years" away, but of the others visible to the naked eye they are too far distant to come within the range of trigonometrical calculation. They are out of reach of the mathematical eye in the depth of space. But we know for certain that the distance of none of these visible stars, without a measurable parallax, is less than four million times the distance of our sun from the earth. It would be useless to express this in figures as it would be altogether incomprehensible. What then can be said of the telescopic stars, not to speak at all of those beyond the power of instruments to determine.

If a railroad could be constructed to the nearest star and the fare made one cent a mile, a single passage would cost $250,000,000,000, that is two hundred and fifty billion dollars, which would make a 94-foot cube of pure gold. All of the coined gold in the world amounts to but $4,000,000,000 (four billion dollars), equal to a gold cube of 24 feet. Therefore it would take sixty times the world's stock of gold to pay the fare of one passenger, at a cent a mile from the earth to Alpha Centauri.

The light from numbers, probably countless numbers,

of stars is so long in coming to us that they could be blotted out of existence and we would remain unconscious of the fact for years, for hundreds of years, for thousands of years, nay to infinity. Thus if *Sirius* were to collide with some other space traveler and be knocked into smithereens as an Irishman would say, we would not know about it for eight years. In fact if all the stars were blotted out and only the sun left we should still behold their light in the heavens and be unconscious of the extinction of even some of the naked-eye stars for sixty or seventy years.

It is vain to pursue farther the unthinkable vastness of the visible Universe; as for the invisible it is equally useless for even imagination to try to grapple with its never-ending immensity, to endeavor to penetrate its awful clouded mystery forever veiled from human view.

In all there are about 3,000 stars visible to the naked eye in each hemisphere. A three-inch pocket telescope brings about one million into view. The grand and scientifically perfected instruments of our great observatories show incalculable multitudes. Every improvement in light-grasping power brings millions of new stars into the range of instrumental vision and shows the "background" of the sky blazing with the light of eye-invisible suns too far away to be separately distinguished.

Great strides are daily being made in stellar photography. Plates are now being attached to the telescopic apparatus whereby luminous heavenly bodies are able to impress their own pictures. Groups of stars are being photographed on one plate. Complete sets of these star photographs are being taken every year, embracing every nook and corner of the celestial sphere and these are carefully compared with one another to find out what changes are going on in the heavens. It will not be long before every star photographically visible to the most powerful telescope will have its present position accurately defined on these photographic charts.

When, the sensitized plate is exposed for a considerable time even invisible stars photograph themselves, and in this way a great number of stars have been discovered which no telescope, however powerful, can bring within the range of vision. Tens of thousands of stars have registered themselves thus on a single plate, and on one occasion an impression was obtained on one plate of more than 400,000.

Astronomers are of the opinion that for every star visible to the naked eye there are more than 50,000 visible to the camera of the telescope. If this is so, then the number of visible stars exceeds 300,000,000 (three hundred millions).

But the picture taking power of the finest photographic lens has a limit; no matter how long the exposure, it cannot penetrate beyond a certain boundary into the vastness of space, and beyond its limits as George Sterling, the Californian poet, says are -

> "fires of unrecorded suns
> That light a heaven not our own."

What is the limit? Answer philosopher, answer sage, answer astronomer, and we have the solution of "the riddle of the Universe."

As yet the riddle still remains, the veil still hangs between the knowable and the unknowable, between the finite and the infinite. Science stands baffled like a wailing creature outside the walls of knowledge importuning for admission. There is little, in truth no hope at all, that she will ever be allowed to enter, survey all the fields of space and set a limit to their boundaries.

Although the riddle of the universe still remains unsolved because unsolvable, no one can deny that Astronomy has made mighty strides forward during the past few years. What has been termed the "Old Astronomy," which concerns itself with the determination of the positions and motions of the heavenly bodies, has been rejuvenated and an immense amount of work has been accomplished by concerted effort, as well as by individual exertions.

The greatest achievements have been the accurate determination of the positions of the fixed stars visible to the eye. Their situation is now estimated with as unerring precision as is that of the planets of our own system. Millions upon millions of stars have been photographed and these photographs will be invaluable in determining the future changes and motions of these giant suns of interstellar space.

Of our own system we now know definitely the laws governing it. Fifty years ago much of our solar machinery was misunderstood and many things were enveloped in mystery which since has been made very plain. The spectroscope has had a wonderful part in astronomical research. It first revealed the nature of the gases existing in the sun. It next enabled us to study the prominences on any clear day. Then by using it in the spectro-heliograph we have been enabled to photograph the entire visible surface of the sun, together with the prominences at one time. Through the spectro-heliograph we know much more about what the central body of our system is doing than our theories can explain. Fresh observations are continually bringing to light new facts which must soon be accounted for by laws at present unknown.

Spectroscopic observations are by no means confined to the sun. By them we now study the composition of the atmospheres of the other planets; through them the presence of chemical elements known on the earth is detected in vagrant comets, far-distant stars and dimly-shining nebulae. The spectroscope also makes it possible to measure the velocities of objects which are approaching or receding from us. For instance we know positively that the bright star called Aldebaran near the constellation of the Pleiades is retreating from us at a rate of almost two thousand miles a minute. The greatest telescopes in the world are now being trained on stars that are rushing away towards the "furthermost" of space and in this way astronomers are trying to get definite knowledge as to the actual velocity with which the celestial bodies are speeding.

It is only within the past few years that photography has been applied to astronomical development. In this connection, more accurate results are obtained by measuring the photographs of stellar spectra than by measuring the spectra themselves. Photography with modern rapid plates gives us, with a given telescope, pictures of objects so faint that no visual telescope of the same size will reveal them. It is in this way that many of the invisible stars have impressed themselves upon exposed plates and given us a vague idea of the immensity in number of those stars which we cannot view with eye or instrument.

Though we have made great advancement, there are many problems yet even in regard to our own little system of sun worlds which clamor loudly for solution. The sun himself represents a crowd of pending problems. His peculiar mode of rotation; the level of sunspots; the constitution of the photospheric cloud-shell, its relation to faculae which rise from it, and to the surmounting vaporous strata; the nature of the prominences; the alternations of coronal types; the affinities of the zodiacal light - all await investigation.

A great telescope has recently shown that one star in eighteen on the average is a visual double - is composed of two suns in slow revolution around their common center of mass. The spectroscope using the photographic plate, has established within the last decade that one star in every five or six on the average is attended by a companion so near to it as to remain invisible in the most powerful telescopes, and so massive as to swing the visible star around in an elliptic orbit.

The photography of comets, nebulae and solar coronas has made the study of these phenomena incomparably more effective than the old visual methods. There is no longer any necessity to make "drawings" of them. The old dread of comets has been relegated into the shade of ignorance. The long switching tails regarded so ominously and from which were anticipated such dire calamities as the destruction of worlds into chaos have been proven to be composed of gaseous vapors of no more solidity than the 'airy nothingness of dreams."

The earth in the circle of its orbit passed through the tail of Halley's comet in May, 1910, and we hadn't even a pyrotechnical display of fire rockets to celebrate the occasion. In fact there was not a single celestial indication of the passage and we would not have known only for the calculations of the astronomer. The passing of a comet now, as far as fear is concerned, means no more, in fact not as much, as the passing of an automobile.

Science no doubt has made wonderful strides in our time, but far as it has gone, it has but opened for us the first few pages of the book of the heavens - the last pages of which no man shall ever read. For aeons upon aeons of time, worlds and suns, and systems of worlds and suns, revolved through the infinity of space, before man made his appearance on the tiny molecule of matter we call the earth, and for aeons upon aeons, for eternity upon eternity, worlds and suns shall continue to roll and revolve after the last vestige of man shall have disappeared, nay after the atoms of earth and sun and all his attending planets of our system shall have amalgamated themselves with other systems in the boundlessness of space; destroyed, obliterated, annihilated, they shall never be, for matter is indestructible. When it passes from one form it enters another; the dead animal that is cast into the earth lives again in the trees and shrubs and flowers and grasses that grow in the earth above where its body was cast. Our earth shall die in course of time, that is, its particles will pass into other compositions and it will be so of the other planets, of the suns, of the stars themselves, for as soon as the old ones die there will ever be new forms to which to attach themselves and thus the process of world development shall go on forever.

The nebulae which astronomers discover throughout the stellar space are extended masses of glowing gases of different forms and are worlds in process of formation. Such was the earth once. These gases solidify and contract and cool off until finally an inhabited world, inhabited by some kind of creatures, takes its place in the whirling galaxy of systems.

The stars which appear to us in a yellow or whitish yellow light are in the heyday of their existence, while those that present a red haze are almost burnt out and will soon become blackened, dead things disintegrating and crumbling and spreading their particles throughout space. It is supposed this little earth of ours has a few more million years to live, so we need not fear for our personal safety while in mortal form.

To us ordinary mortals the mystery as well as the majesty of the heavens have the same wonderful attraction as they had for the first of our race. Thousands of years ago the black-bearded shepherds of Eastern lands gazed nightly into the vaulted dome and were struck with awe as well as wonder in the contemplation of the glittering specks which appeared no larger than the pebbles beneath their feet.

We in our time as we gaze with unaided eye up at the mighty disk of the so called Milky Way, no longer regard the scintillating points glittering like diamonds in a jeweler's show-case, with feelings of awe, but the wonder is still upon us, wonder at the immensity of the works of Him who built the earth and sky, who, "throned in height sublime, sits amid the cherubim," King of the Universe, King of kings and Lord of lords. With a deep faith we look up and adore, then reverently exclaim, - "Lord, God! wonderful are the works of Thy Hands."

CHAPTER XV

CAN WE COMMUNICATE WITH OTHER WORLDS?

Vastness of Nature - Star Distances - Problem of Communicating with Mars - The Great Beyond.

A story is told of a young lady who had just graduated from boarding school with high honors. Coming home in great glee, she cast her books aside as she announced to her friends; - "Thank goodness it is all over, I have nothing more to learn. I know Latin and Greek, French and German, Spanish and Italian; I have gone through Algebra, Geometry, Trigonometry, Conic Sections and the Calculus; I can interpret Beethoven and Wagner, and - but why enumerate? - in short, '*I know everything.*'"

As she was thus proclaiming her knowledge her hoary-headed grandfather, a man whom the Universities of the world had honored by affixing a score of alphabetical letters to his name, was experimenting in his laboratory. The lines of long and deep study had corrugated his brow and furrowed his face. Wearily he bent over his retorts and test tubes. At length he turned away with a heavy sigh, threw up his hands and despairingly exclaimed, - "Alas, alas! after fifty years of study and investigation, I find *I know nothing.*"

There is a moral in this story that he who runs may read. Most of us are like the young lady, - in the pride of our ignorance, we fancy we know almost everything. We boast of the progress of our time, of what has been accomplished in our modern world, we proclaim our triumphs from the hilltops, - "Ha!" we shout, "we have annihilated time and distance; we have conquered the forces of nature and made them subservient to our will; we have chained the lightning and imprisoned the thunder; we have wandered through the fields of space and measured the dimensions and revolutions of stars and suns and planets and systems. We have opened the eternal gates of knowledge for all to enter and crowned man king of the universe."

Vain boasting! The gates of knowledge have been opened, but we have merely got a peep at what lies within. And man, so far from being king of the universe, is but as a speck on the fly-wheel that controls the mighty machinery of creation. What we know is infinitesimal to what we do not know. We have delved in the fields of science, but as yet our ploughshares have merely scratched the tiniest portion of the surface, - the furrow that lies in the distance is unending. In the infinite book of knowledge we have just turned over a few of the first pages; but as it is infinite, alas! we can never hope to reach the final page, for there is no final page. What we have accomplished is but as a mere drop in the ocean, whose waves wash the continents of eternity. No scholar, no scientist can bound those continents, can tell the limits to which they stretch, inasmuch as they are illimitable.

Ask the most learned *savant* if he can fix the boundaries of space, and he will answer, - No! Ask him if he can define *mind* and *matter*, and you will receive the same answer.

"What is mind? It is no matter."

"What is matter? Never mind."

The atom formerly thought to be indivisible and the smallest particle of matter has been reduced to molecules, corpuscles, ions, and electrons; but the nature, the primal cause of these, the greatest scientists on earth are unable to determine. Learning is as helpless as ignorance when brought up against this stone-wall of mystery. *The effect* is seen, but the *cause* remains indeterminable. The scientist, gray-haired in experience and experiment, knows no more in this regard than the prattling child at its mother's knee. The child asks, - "Who made the world?" and the mother answers, "God made the world." The infant mind, suggestive of the future craving for knowledge, immediately asks, - "Who is God?" Question of questions to which the philosopher and the peasant must give the same answer, - "God is the infinite, the eternal, the source of all things, the *alpha* and *omega* of creation, from Him all came, to Him all must return." He is the beginning of Science, the foundation on which our edifice of knowledge rests.

We hear of the conflict between Science and Religion. There is no conflict, can be none, for all Science must be based on faith, - faith in Him who holds worlds and suns "in the hollow of His hand." All our great scientists have been deeply religious men, acknowledging their own insignificance before Him who fills the universe with His presence.

What is the universe and what place do we hold in it? The mind of man becomes appalled in consideration of the question. The orb we know as the sun is centre of a system of worlds of which our earth is almost the most insignificant; yet great as is the sun when compared to the little bit of matter on which we dwell and have our being, it is itself but a mote, as it were, in the beam of the Universe. Formerly this sun was thought to be fixed and immovable, but the progress of science demonstrated that while the earth moves around this luminary, the latter is moving with mighty velocity in an orbit of its own. Tis the same with all the other bodies which we erroneously call "fixed stars." These stars are the suns of other systems of worlds, countless systems, all rushing through the immensity of space, for there is nothing fixed or stationary in creation, - all is movement, constant, unvarying. Suns and stars and systems perform their revolutions with unerring precision, each unit-world true to its own course, thus proving to the soul of reason and the consciousness of faith that there must needs be an omnipotent hand at the lever of this grand machinery of the universe, the hand that fashioned it, that of God. Addison beautifully expresses the idea in referring to the revolutions of the stars:

> "In reason's ear they all rejoice,
> And utter forth one glorious voice,
> Forever singing as they shine-
> 'The Hand that made us is Divine.'"

Our sun, the centre of the small system of worlds of which the earth is one, is distant from us about ninety-three million miles. In winter it is nearer; in summer farther off. Light travels this distance in about eight minutes, to be exact, the rate is 186,400 miles per second. To get an idea of the immensity of the distance of the so-called fixed stars, let us take this as a base of comparison. The nearest fixed star to us is *Alpha Centauri*, which is one of the brightest as seen in the southern heavens. It requires four and one-quarter years for a beam of light to travel from this star to earth at the rate of 186,000 miles a second, thus showing that Alpha Centauri is about two hundred and seventy-five thousand times as far from us as is the sun, in other words, more than 25,575,000,000,000 miles, which, expressed in our notation, reads twenty-five trillion, five hundred and seventy-five billion miles, a number which the mind of man is incapable of grasping. To use the old familiar illustration of the express train, it would take the "Twentieth Century Limited," which does the thousand mile trip between New York and Chicago in less than twenty-four hours, some one million two hundred and fifty thousand years at the same speed to travel from the earth to *Alpha Centauri*. *Sirius*, the Dog-Star, is twice as far away, something like eight or nine "light" years from our solar system; the Pole-Star is forty-eight "light" years removed from us, and so on with the rest, to an infinity of numbers. From the dawn of creation in the eternal cosmos of matter, light has been travelling from some stars in the infinitude of space at the rate of 186,000 miles per second, but so remote are they from our system that it has not reached us as yet. The contemplation is bewildering; the mind sinks into nothingness in consideration of a magnitude so great and distance so confusing. What lies beyond? - a region which numbers cannot measure and thought cannot span, and beyond that? - the eternal answer, - GOD.

In face of the contemplation of the vastness of creation, of its boundlessness the question ever obtrudes itself, - What place have we mortals in the universal cosmos? What place have we finite creatures, who inhabit this speck of matter we call the earth, in this mighty scheme of suns and systems and never-ending space. Does the Creator of all think us the most important of his works, that we should be the particular objects of revelation, that for us especially heaven was built, and a God-man, the Son of the Eternal, came down to take flesh of our flesh and live among us, to show us the way, and finally to offer himself as a victim to the Father to expiate our transgressions. Mystery of mysteries before which we stand appalled and lost in wonder. Self-styled rationalists love to point out the irrationality and absurdity of supposing that the Creator of all the unimaginable vastness of suns and systems, filling for all we know endless space, should take any special interest in so mean and pitiful a creature as man, inhabiting such an infinitesimal speck of matter as the earth, which depends for its very life and light upon a second or third-rate or hundred-rate Sun.

From the earliest times of our era, the sneers and taunts of atheism and agnosticism have been directed at the humble believer, who bows down in submission and questions not. The fathers of the Church, such as Augustine and Chrysostom and Thomas of Aquinas and, at a later time, Luther, and Calvin, and Knox, and Newman, despite the war of creeds, have attacked the citadel of the scoffers; but still the latter hurl their javelins from the ramparts, battlements and parapets and refuse to be repulsed. If there are myriads of other worlds, thousands, millions of them in point of magnitude greater than ours, what concern say they has the Creator with our little atom of matter? Are other worlds inhabited besides our own. This is the question that will not down - that is always begging for an answer. The most learned savants of modern time, scholars, sages, philosophers and scientists have given it their attention, but as yet no one has been able to conclusively decide whether a race of intelligent beings exists in any sphere other than our own. All efforts to determine the matter result in mere surmise, conjecture and guesswork. The best of scientists can only put forward an opinion.

Professor Simon Newcomb, one of the most brilliant minds our country has produced, says: "It is perfectly reasonable to suppose that beings, not only animated but endowed with reason, inhabit countless worlds in space." Professor Mitchell of the Cincinnati Observatory, in his work, "Popular Astronomy," says, - "It is most incredible to assert, as so many do, that our planet, so small and insignificant in its proportions when compared with the planets with which it is allied, is the only world in the whole universe filled with sentient, rational, and intelligent beings capable of comprehending the grand mysteries of the physical universe." Camille Flammarion, in referring to the utter insignificance of the earth in the immensity of space, puts forward his view thus: "If advancing with the velocity of light we could traverse from century to century the unlimited number of suns and spheres without ever meeting any limit to the prodigious immensity where God brings forth his worlds, and looking behind, knowing not in what part of the infinite was the little grain of dust called the earth, we would be compelled to unite our voices with that universal nature and exclaim - 'Almighty God, how senseless were we to believe that there was nothing beyond the earth and that our abode alone possessed the privilege of reflecting Thy greatness and honor.'"

The most distinguished astronomers and scientists of a past time, as well as many of the most famous divines, supported the contention of world life beyond the earth. Among these may be mentioned Kepler and Tycho, Giordano Bruno and Cardinal Cusa, Sir William and Sir John Herschel, Dr. Bentley and Dr. Chalmers, and even Newton himself subscribed in great measure to the belief that the planets and stars are inhabited by intelligent beings.

Those who deny the possibility of other worlds being inhabited, endeavor to show that our position in the universe is unique, that our solar system is quite different from all others, and, to crown the argument, they assert that our little world has just the right amount of water, air, and gravitational force to enable it to be the abode of intelligent life, whereas elsewhere, such conditions do not prevail, and that on no other sphere can such physical habitudes be found as will enable life to originate or to exist. It can be easily shown that such reasoning is based on untenable foundations. Other worlds have to go through processes of evolution, and there can be no doubt that many are in a state similar to our own. It required hundreds of thousands, perhaps hundreds of millions of years, before this earth was fit to sustain human life. The same transitions which took place on earth are taking place in other planets of our system, and other systems, and it is but reasonable to assume that in other systems there are much older worlds than the earth, and that these have arrived at a more developed state of existence, and therefore have a life much higher than our own. As far as physical conditions are concerned, there are suns similar to our own, as revealed by the spectroscope, and which have the same eruptive energy. Astronomical Science has incontrovertibly demonstrated, and evidence is continually increasing to show that dark, opaque worlds like ours exist and revolve around their primaries. Why should not these worlds be inhabited by a race equal or even superior in intelligence to ourselves, according to their place in the cosmos of creation?

Leaving out of the question the outlying worlds of space, let us come to a consideration of the nearest celestial neighbor we have in our own system, the planet Mars: Is there rational life on Mars and if so can we communicate with the inhabitants?

Though little more than half the earth's size, Mars has a significance in the public eye which places it first in importance among the planets. It is our nearest neighbor on the outer side of the earth's path around the Sun and, viewed through a telescope of good magnifying power, shows surface markings, suggestive of continents, mountains, valleys, oceans, seas and rivers, and all the varying phenomena which the mind associates with a world like unto our own. Indeed, it possesses so many features in common with the earth, that it is impossible to resist the conception of its being inhabited. This, however, is not tantamount to saying that if there is a race of beings on Mars they are the same as we on Earth. By no means. Whatever atmosphere exists on Mars must be much thinner than ours and far too rare to sustain the life of a people with our limited lung capacity. A race with immense chests could live under such conditions, and folk with gills like fish could pass a comfortable existence in the rarefied air. Besides the tenuity of the atmosphere, there are other conditions which would cause life to be much different on Mars. Attraction and gravitation are altogether different. The force with which a substance is attracted to the surface of Mars is only a little more than one-third as strong as on the earth. For instance one hundred pounds on Earth would weigh only about thirty-eight pounds on Mars. A man who could jump five feet here could clear fifteen feet on Mars. Paradoxical as it may seem, the smaller a planet, in comparison with ours and consequently the less the pull of gravity at its centre, the greater is the probability that its inhabitants, if any, are giants when compared with us. Professor Lowell has pointed out that to place the Martians (if there are such beings) under the same conditions as those in which we exist, the average inhabitant must be considered to be three times as large and three times as heavy as the average human being; and the strength of the Martians must exceed ours to even a greater extent than the bulk and weight; for their muscles would be twenty-seven times more effective. In fact, one Martian could do the work of fifty or sixty men.

It is idle, however, to speculate as to what the forms of life are like on Mars, for if there are any such forms our ideas and conceptions of them must be imaginary, as we cannot see them on Mars we do not know. There is yet no possibility of seeing anything on the planet less than thirty miles across, and even a city of that size, viewed through the most powerful telescope, would only be visible as a minute speck. Great as is the perfection to which our optical instruments have been brought, they have revealed nothing on the planet save the so-called canals, to indicate the presence of sentient rational beings. The canals discovered by Schiaparelli of the Milan Observatory in 1877 are so regular, outlined with such remarkable geometrical precision, that it is claimed they must be artificial and the work of a high order of intelligence. "The evidence of such work," says Professor Lowell, "points to a highly intelligent mind behind it."

Can this intelligence in any way reach us, or can we express ourselves to it? Can the chasm of space which lies between the Earth and Mars be bridged - a chasm which, at the shortest, is more than thirty-five million miles across or one hundred and fifty times greater than the distance between the earth and the moon? Can the inhabitants of the Earth and Mars exchange signals? To answer the question, let us institute some comparisons. Suppose the fabled "Man in the Moon" were a real personage, we would require a telescope 800 times more powerful than the finest instrument we now have to see him, for the space penetrating power of the best telescope is not more than 300 miles and the moon is 240,000 miles distant. An object to be visible on the moon would require to be as large as the Metropolitan Insurance Building in New York, which is over 700 feet high. To see, therefore, an object on Mars by means of the telescope the object would need to have dimensions one hundred and fifty times as great as the object on the moon; in other words, before we could see a building on Mars, it would have to be one hundred and fifty times the size of the Metropolitan Building. Even if there are inhabitants there, it is not likely they have such large buildings.

Assuming that there *are* Martians, and that they are desirous of communicating with the earth by waving a flag, such a flag in order to be seen through the most powerful telescopes and when Mars is nearest, would have to be 300 miles long and 200 miles wide and be flung from a flagpole 500 miles high. The consideration of such a signal only belongs to the domain of the imagination. As an illustration, it should conclusively settle the question of the possibility or rather impossibility of signalling between the two planets.

Let us suppose that the signalling power of wireless telegraphy had been advanced to such perfection that it was possible to transmit a signal across a distance of 8,000 miles, equal to the diameter of the earth, or 1-30 the distance to the moon. Now, in order to be appreciable at the moon it would require the intensity of the 8,000 mile ether waves to be raised not merely 30 times, but 30 times 30, for to use the ordinary expression, the intensity of an effect spreading in all directions like the ether waves, decreases inversely as the square of the distance. If the whole earth were brought within the domain of wireless telegraphy, the system would still have to be improved 900 times as much again before the moon could be brought within the sphere of its influence. A wireless telegraphic signal, transmitted across a distance equal to the diameter of the earth, would be reduced to a mere sixteen-millionth part if it had to travel over the distance to Mars; in other words, if wireless telegraphy attained the utmost excellence now hoped for it - that is, of being able to girdle the earth - it would have to be increased a thousandfold and then a thousandfold again, and finally multiplied by 16, before an appreciable *signal* could be transmitted to Mars. This seems like drawing the long bow, but it is a scientific truth. There is no doubt that ether waves can and do traverse the distance between the Earth and Mars, for the fact that sunlight reaches Mars and is reflected back to us proves this; but the source of waves adequate to accomplish such a feat must be on such a scale as to be hopelessly beyond the power of man to initiate or control. Electrical signalling to Mars is much more out of the question than wireless. Even though electrical phenomena produced in any one place were sufficiently intense to be appreciable by suitable instruments all over the earth, that intensity would have to be enhanced another sixteen million-fold before they would be appreciable on the planet Mars.

It is absolutely hopeless to try to span the bridge that lies between us and Mars by any methods known to present day science. Yet men styling themselves scientists say it can be done and will be done. This is a prophecy, however, which must lie in the future.

As has been pointed out, we have as yet but scratched the outer surface in the fields of knowledge. What visions may not be opened to the eyes of men, as they go down deeper and deeper into the soil. Secrets will be exhumed undreamt of now, mysteries will be laid bare to the light of day, and perhaps the psychic riddle of life itself may be solved. Then indeed, Mars may come to be looked on as a next-door neighbor, with whose life and actions we are as well acquainted as with our own. The thirty-five million miles that separate him from us may be regarded as a mere step in space and the most distant planets of our system as but a little journey afield. Distant Uranus may be looked upon as no farther away than is, say, Australia from America at the present time.

It is vain, however, to indulge in these premises. The veil of mystery still hangs between us and suns and stars and systems. One fact lies before us of which there is no uncertainty - *we die* and pass away from our present state into some other. We are not annihilated into nothingness. Suns and worlds also die, after performing their allotted revolutions in the cycle of the universe. Suns glow for a time, and planets bear their fruitage of plants and animals and men, then turn for aeons into a dreary, icy listlessness and finally crumble to dust, their atoms joining other worlds in the indestructibility of matter.

After all, there really is no death, simply change - change from one state to another. When we say we die, we simply mean that we change our state. There is a life beyond the grave. As Longfellow beautifully expresses it:

> "Life is real, life is earnest,
> And the grave is not its goal,
> Dust thou art, to dust returnest,
> Was not spoken of the soul."

But whither do we go when we pass on? Where is the soul when it leaves the earthly tenement called the body? We, Christians, in the light of revelation and of faith, believe in a heaven for the good; but it is not a material place, only a state of being. Where and under what conditions is that state? This leads us to the consideration of another question which is engrossing the minds of many thinkers and reasoners of the present day. Can we communicate with the Spirit world? Despite the tenets and beliefs and experiences of learned and sincere investigators, we are constrained, thus far, to answer in the negative.

Yet, though we cannot communicate with it, we know there is a spirit world; the inner consciousness of our being apprises us of that fact, we know our loved ones who have passed on are not dead but gone before, just a little space, and that soon we shall follow them into a higher existence. As Talmage said, the tombstone is not the terminus, but the starting post, the door to the higher life, the entrance to the state of endless labor, grand possibilities, and eternal progression.

Choose from Thousands of 1stWorldLibrary Classics By

Adolphus WilliamWard	Clemence Housman	Gabrielle E. Jackson
Aesop	Confucius	Garrett P. Serviss
Agatha Christie	Cornelis DeWitt Wilcox	Gaston Leroux
Alexander Aaronsohn	Cyril Burleigh	George Ade
Alexander Kielland	D. H. Lawrence	Geroge Bernard Shaw
Alexandre Dumas	Daniel Defoe	George Ebers
Alfred Gatty	David Garnett	George Eliot
Alfred Ollivant	Don Carlos Janes	George MacDonald
Alice Duer Miller	Donald Keyhole	George Orwell
Alice Turner Curtis	Dorothy Kilner	George Tucker
Alice Dunbar	Dougan Clark	George W. Cable
Ambrose Bierce	E. Nesbit	George Wharton James
Amelia E. Barr	E.P.Roe	Gertrude Atherton
Andrew Lang	E. Phillips Oppenheim	Grace E. King
Andrew McFarland Davis	Edgar Allan Poe	Grant Allen
Anna Sewell	Edgar Rice Burroughs	Guillermo A. Sherwell
Annie Besant	Edith Wharton	Gulielma Zollinger
Annie Hamilton Donnell	Edward J. O'Biren	Gustav Flaubert
Annie Payson Call	John Cournos	H. A. Cody
Anton Chekhov	Edwin L. Arnold	H. B. Irving
Arnold Bennett	Eleanor Atkins	H. G. Wells
Arthur Conan Doyle	Elizabeth Cleghorn Gaskell	H. H. Munro
Arthur Ransome		H. Irving Hancock
Atticus	Elizabeth Von Arnim	H. Rider Haggard
B. M. Bower	Ellem Key	H. W. C. Davis
Basil King	Emily Dickinson	Hamilton Wright Mabie
Bayard Taylor	Erasmus W. Jones	Hans Christian Andersen
Ben Macomber	Ernie Howard Pie	Harold Avery
Booth Tarkington	Ethel Turner	Harold McGrath
Bram Stoker	Ethel Watts Mumford	Harriet Beecher Stowe
C. Collodi	Eugenie Foa	Harry Houidini
C. E. Orr	Eugene Wood	Helent Hunt Jackson
C. M. Ingleby	Evelyn Everett-Green	Helen Nicolay
Carolyn Wells	Everard Cotes	Hendy David Thoreau
Catherine Parr Traill	F. J. Cross	Henrik Ibsen
Charles A. Eastman	Federick Austin Ogg	Henry Adams
Charles Dickens	Ferdinand Ossendowski	Henry Ford
Charles Dudley Warner	Francis Bacon	Henry Frost
Charles Farrar Browne	Francis Darwin	Henry James
Charles Ives	Frances Hodgson Burnett	Henry Jones Ford
Charles Kingsley	Frank Gee Patchin	Henry Seton Merriman
Charles Lathrop Pack	Frank Harris	Henry Wadsworth Longfellow
Charles Whibley	Frank Jewett Mather	
Charles Willing Beale	Frank L. Packard	Henry W Longfellow
Charlotte M. Braeme	Frederick Trevor Hill	Herbert A. Giles
Charlotte M.Yonge	Frederick Winslow Taylor	Herbert N. Casson
Clair W. Hayes	Friedrich Kerst	Herman Hesse
Clarence Day Jr.	Friedrich Nietzsche	Homer
Clarence E. Mulford	Fyodor Dostoyevsky	Honore De Balzac

Horace Walpole	Laurence Housman	Robert Lansing
Horatio Alger, Jr.	Leo Tolstoy	Robert Michael Ballantyne
Howard Pyle	Leonid Andreyev	Robert W. Chambers
Howard R. Garis	Lewis Carroll	Rosa Nouchette Carey
Hugh Lofting	Lilian Bell	Ross Kay
Hugh Walpole	Lloyd Osbourne	Rudyard Kipling
Humphry Ward	Louis Tracy	Samuel B. Allison
Ian Maclaren	Louisa May Alcott	Samuel Hopkins Adams
Israel Abrahams	Lucy Fitch Perkins	Sarah Bernhardt
J.G.Austin	Lucy Maud Montgomery	Selma Lagerlof
J. Henri Fabre	Lydia Miller Middleton	Sherwood Anderson
J. M. Barrie	Lyndon Orr	Sigmund Freud
J. Macdonald Oxley	M. H. Adams	Standish O'Grady
J. S. Knowles	Margaret E. Sangster	Stanley Weyman
J. Storer Clouston	Margaret Vandercook	Stella Benson
Jack London	Maria Edgeworth	Stephen Crane
Jacob Abbott	Maria Thompson Daviess	Stewart Edward White
James Allen	Mariano Azuela	Stijn Streuvels
James Lane Allen	Marion Polk Angellotti	Swami Abhedananda
James Andrews	Mark Overton	Swami Parmananda
James Baldwin	Mark Twain	T. S. Ackland
James DeMille	Mary Austin	The Princess Der Ling
James Joyce	Mary Cole	Thomas A. Janvier
James Oliver Curwood	Mary Rowlandson	Thomas A Kempis
James Oppenheim	Mary Wollstonecraft	Thomas Anderton
James Otis	Shelley	Thomas Bailey Aldrich
Jane Austen	Max Beerbohm	Thomas Bulfinch
Jens Peter Jacobsen	Myra Kelly	Thomas De Quincey
Jerome K. Jerome	Nathaniel Hawthrone	Thomas H. Huxley
John Burroughs	O. F. Walton	Thomas Hardy
John F. Kennedy	Oscar Wilde	Thomas More
John Gay	Owen Johnson	Thornton W. Burgess
John Glasworthy	P.G.Wodehouse	U. S. Grant
John Habberton	Paul and Mable Thorn	Valentine Williams
John Joy Bell	Paul G. Tomlinson	Victor Appleton
John Milton	Paul Severing	Virginia Woolf
John Philip Sousa	Peter B. Kyne	Walter Scott
Jonathan Swift	Plato	Washington Irving
Joseph Carey	R. Derby Holmes	Wilbur Lawton
Joseph Conrad	R. L. Stevenson	Wilkie Collins
Joseph Jacobs	Rabindranath Tagore	Willa Cather
Julian Hawthrone	Rahul Alvares	Willard F. Baker
Julies Vernes	Ralph Waldo Emmerson	William Makepeace Thackeray
Justin Huntly McCarthy	Rene Descartes	
Kakuzo Okakura	Rex E. Beach	William W. Walter
Kenneth Grahame	Richard Harding Davis	Winston Churchill
Kate Langley Bosher	Richard Jefferies	Yei Theodora Ozaki
L. A. Abbot	Robert Barr	Young E. Allison
L. T. Meade	Robert Frost	Zane Grey
L. Frank Baum	Robert Gordon Anderson	
Laura Lee Hope	Robert L. Drake	

www.ingramcontent.com/pod-product-compliance
Lightning Source LLC
Chambersburg PA
CBHW022359040426
42450CB00005B/255